Probing Polymer Structures

Probing Polymer Structures

Jack L. Koenig, EDITOR
Case Western Reserve University

Based on a symposium
cosponsored by the Divisions
of Polymer Chemistry and
Analytical Chemistry at the
174th Meeting of the
American Chemical Society,
Chicago, Illinois,
August 29–September 2, 1977.

ADVANCES IN CHEMISTRY SERIES **174**

AMERICAN CHEMICAL SOCIETY
WASHINGTON, D. C. 1979

Library of Congress CIP Data

Probing polymer structures.
(Advances in chemistry series; 174 ISSN 0065-2393)

Includes bibliographies and index.

1. Polymers and polymerization—Analysis—Congresses.
I. Koenig, Jack L. II. American Chemical Society. Division of Polymer Chemistry. III. American Chemical Society. Division of Analytical Chemistry. IV. Series.

QD1.A355 no. 174 [QD380] 540'.8s [547'.84]
ISBN 0-8412-0406-3 79-10008
ADCSAJ 174 1-277 1979

PRINTED IN THE UNITED STATES OF AMERICA

FOREWORD

Advances in Chemistry Series was founded in 1949 by the American Chemical Society as an outlet for symposia and collections of data in special areas of topical interest that could not be accommodated in the Society's journals. It provides a medium for symposia that would otherwise be fragmented, their papers distributed among several journals or not published at all. Papers are reviewed critically according to ACS editorial standards and receive the careful attention and processing characteristic of ACS publications. Volumes in the Advances in Chemistry Series maintain the integrity of the symposia on which they are based; however, verbatim reproductions of previously published papers are not accepted. Papers may include reports of research as well as reviews since symposia may embrace both types of presentation.

CONTENTS

PREFACE

The goal of polymer research is to develop improved performance through an understanding of the relationships between structure and properties. In the past decade it has become clear that most of these improvements will have to be made on commercial polymers which are presently being produced. Consequently, numerous sophisticated techniques and instruments have been developed in order to detect, characterize, and monitor the structure and properties of commercial polymers. This book represents an aid to the polymer scientist in choosing from among the instrumentation and techniques available to help him understand the microstructural variables of polymers, and to bring him up-to-date regarding instrumentation, developing techniques, and new applications.

This book presents current active areas of research in the instrumental methods of characterizing polymers. It is intended to represent a blend of the theoretical and experimental bases of the various techniques and the recent applications of these methods in an effort to develop a fundamental understanding of the manner in which polymers perform in commercial systems. Virtually every physical phenomena from noise associated with flow or crazing to scattering of neutrons, photons, electrons, and x-rays, have been used in some fashion to probe polymeric systems. This book will outline the use of these phenomena for instrumentally studying the behavior of polymers. Initially, the processes of flow, crazing, and deformation will be probed using the measurements of sound, stress, and fragmentation products. A sensitive tensile testing instrument, dubbed the nanotensilometer, will be described. This apparatus allows the measurement of stress at the level of dynes and strain at the dimensions equivalent to the smallest morphological unit—the single crystal. The degradation products produced by pyrolysis, thermal degradation, or stress, which are detected by IR spectroscopy or mass spectroscopy, allow determination of the structure of the initial polymer chain and its modes of breakdown. The scattering of photons and neutrons by internal molecular motion or by external molecular diffusion processes can be detected by quasieleastic light scattering techniques or with a multipass Fabry-Perot spectrometer and will be described herein. These measurements give us new insights into the size, shape, and motion of polymer molecules in solution and in the solid state. Of course, the diffusion process itself can be used as a molecular probe and this method of structural analysis is described. The classical analysis techniques such as vibrational spectroscopy and nuclear magnetic resonance have received sufficient attention in the literature that their chemical applications will not be discussed except as to their role in studying transition and relaxation processes. A novel new method of obtaining vibrational energy level information using inelastic electron tunneling is described and applied to chemisorbed substances. Also, new applications of electron spectroscopy to the study of structure and bonding in polymers are given attention.

The papers that comprise this volume were presented originally at a symposium cosponsored by the Polymer and Analytical Chemistry Divisions of the American Chemical Society. The papers have been expanded and updated for the present volume and I wish to thank the contributors for their cooperation in this project. The support of the ACS Divisions of Polymer Chemistry and Analytical Chemistry is also acknowledged.

Department of Macromolecular Science
Case Western Reserve University
Cleveland, Ohio 44106
February 9, 1979

JACK L. KOENIG

1

Noise Phenomena Associated with Thermal Transitions and Flow in Certain Polymeric Systems

C. KLASON and J. KUBÁT

Department of Polymeric Materials, Chalmers University of Technology, S-40220 Gothenburg, Sweden

Measurements of thermal and current noise showed pronounced noise maxima slightly above the glass transition (T_g) *and melting point* (T_m) *of carbon-black-filled polymers. The samples, polystyrene and HD-polyethylene, were rendered conductive by adding minor amounts of carbon black. A strong time dependence on the noise level was recorded around the* T_g *peak. The peaks are probably associated with changes in the carbon black network at the critical temperatures. Aqueous solutions of polyethylene oxide, hydroxyethyl cellulose, carboxymethyl cellulose, polyvinylpyrrolidone, and polyvinyl alcohol showed an increase in current noise with the shear rate during Couette flow. A simple relationship between the value of* α *in the* $1/f^{\alpha}$ *noise spectra and the exponent* n *of the Ostwald–de Waele law (degree of pseudoplasticity) was found.*

This chapter summarizes the results of a preliminary investigation concerning the possibility of gaining new information about the mechanisms of transitions and flow in liquid and solid polymeric systems from measurements of the accompanying changes in electrical noise. With regard to current noise, the idea behind such an approach appears especially simple. In such a case, a current flowing through the sample is expected to be modulated by processes taking place on the molecular or larger level. An analysis of the spectral characteristics of the modulated noise could thus produce new information about the nature of the

0-8412-0406-3/79/33-174-001$05.00/1

microprocesses occuring during a thermal transition or flow. By studying the characteristics of thermal noise possible deviations from thermal equilibrium could be detected.

An investigation of thermal and current noise under conditions of flow could be rewarding because the physical mechanisms underlying flow in polymeric systems are not especially well understood. To a certain degree, this applies also to thermally induced transitions including the glass transition.

The results presented here illustrate the general feasibility of this technique. They relate primarily to the behavior of thermal and current noise in the glass transition (T_g) or melting (T_m) region of an amorphous (polystyrene) and a crystalline (HD-polyethylene (HDPE)) polymer rendered conductive by the addition of minor amounts of carbon black, and further they relate to the noise of aqueous solutions of certain polymers during Couette flow. Because of experimental difficulties, noise measurements on solid polymers during deformation and flow have not yet produced useful results.

When measuring thermal noise of carbon-black-filled PS and other amorphous polymers pronounced maxima were found near the vicinity of T_g, which are very similar to the maxima in resistivities reported earlier (*1*). The intensity of these peaks depended on the rate of cooling/heating of the sample prior to and during the measurements and also on the storage time. With HDPE, noise peaks were recorded in the vicinity of T_m, and the time dependence of the peaks was less pronounced.

By relating corresponding noise and resistivity values, it could be shown that deviations from thermodynamic equilibrium did not occur in the frequency range covered, i.e., $10–10^4$ Hz. The results relating to current noise were largely similar.

The results obtained during the Couette flow of aqueous solutions of polyethylene oxide and other water-soluble polymers appear especially promising since they showed an appreciable increase in the current noise level with shear rate. The current noise level depended also on the viscosity (molecular weight) of the solution. A slight increase of thermal noise was recorded also. The pseudoplasticity exponent n in the Ostwald–de Waele power law formula and the exponent α in the $1/f^{\alpha}$-frequency distribution of the current noise were interrelated. This relation appeared to be generally valid.

Definitions

In order to clarify some basic concepts regarding electrical fluctuations the following definitions are necessary.

Thermal noise denotes electrical fluctuations spontaneously generated in a sample resulting from thermal agitation. The noise level is given by the Nyquist formula (*2*):

$$E^2 = 4\,k\,T\,R\,B \tag{1}$$

where E is the noise voltage, T is the temperature, R is the resistance of the sample within the bandwidth B, and k is the Bolzmann constant. This type of noise is termed white, implying that its spectral intensity is constant provided R is independent of frequency. For example, within a bandwidth of 10^4 Hz a 1-MΩ resistor generates a fluctuation voltage of 13 μV at room temperature.

Current noise is the noise component exceeding the thermal noise level. In systems relevant to the present context, i.e., carbon-black-filled polymers, carbon resistors, solutions etc., it normally has a frequency distribution of the form $1/f^{\alpha}$, where α = 1–3. The intensity of the noise thus falls rapidly with increasing frequency. The current noise level usually greatly exceeds that of the thermal noise.

Experimental

Carbon-Black-Filled Polymers. The polymers used were laboratory grades of HDPE (S.N.P.A. Aquitaine, France, $\overline{M}_w$ = 70,300, $\overline{M}_w/\overline{M}_n$ = 1.10) and PS (Polyscience, USA, $\overline{M}_w$ = 227,000, $\overline{M}_w/\overline{M}_n$ = 3.07) containing a low percentage (6.5% and 4%, respectively) of very fine grade carbon black (Ketjenblack EC, BET 800 m^2/g, ASTM iodine number 950 mg/g, DBP 340 mL/100 g, ash 0.7%, and volatile matter 3.5%). The melting point of the HDPE was 135°C, and the T_g of the PS 100°C as determined by the DSC method (Perkin–Elmer DSC-2, heating rate 1°C/min). The carbon black was incorporated into the samples by calendering between heated rolls (Rapra Micromill) until a good dispersion was achieved (over 95% according to the Leigh–Dugmore rating). The samples were given their final disc-shaped form (diameter, 13 mm; thickness, 1 mm) using a heated KBr pellet press (150°C, 200 MPa). Silver paint (Acheson type 1415) served as electrodes. The volume resistivity of the samples was approximately 10^7 ohmcm. Before and during the experiments the samples were stored in a desiccator over P_2O_5. Several other sample preparation methods and electrode materials were used also, but this did not influence the results.

The samples were placed in a shielded thermostatically controlled measuring device. The sample temperature was measured (± 1°C) with a thermocouple attached to a dummy sample placed close to the sample under investigation to reduce spurious radio frequency signals otherwise induced by the thermocouple wires.

The two electrodes attached to the sample were connected to a coaxial terminal of BNC type. The wires and the cable had to be kept short to reduce the capacitive load on the sample. This capacitive loading together with the resistance of the sample acts as a filter to reduce the high-frequency components of the signal. For this reason the sample resistance was reduced by adding carbon black.

During the current noise measurements a constant current was sent through the sample via a wire-wound resistor whose resistance was at least one order of magnitude higher than that of the sample. This wire wound resistor did not produce any additional current noise, and it prevented the noise signal from being loaded by the current supply.

Aqueous Polymer Solutions. The samples used were polyethylene oxide (PEO) (Union Carbide Polyox, WSR N-10, WSR 205, WSR 301, mol wt 10^5, $6 \cdot 10^5$, and $4 \cdot 10^6$, respectively), hydroxyethyl cellulose (Union Carbide type QP100M and QP52000, $\overline{M}_w$ $3.8 \cdot 10^5$ and $3.4 \cdot 10^5$, respectively), sodium carboxymethyl cellulose (Hercules type 7HF; $\overline{M}_w$, $7 \cdot 10^5$), polyvinylpyrrolidone (GAF K90, $\overline{M}_w$ $3.6 \cdot 10^5$) and poly(vinyl alcohol) (PVA) (Hoechst Mowiol N90-98, $\overline{M}_w$ $4.2 \cdot 10^5$). The solutions were prepared by slowing stirring the polymer with distilled and deionized water (conductivity 0.8 μmho/cm). Small amounts of KCl were added to the solution to adjust the conductivity to approximately 10^{-4} mho/cm (to match the measuring amplifier impedance). The apparent viscosity of the solution was of the order of 100 cP at a shear rate of 100 sec^{-1}. The flow curves of the solutions were obtained with a Contraves Epprecht Rheomat using a plate and cone system (23°C).

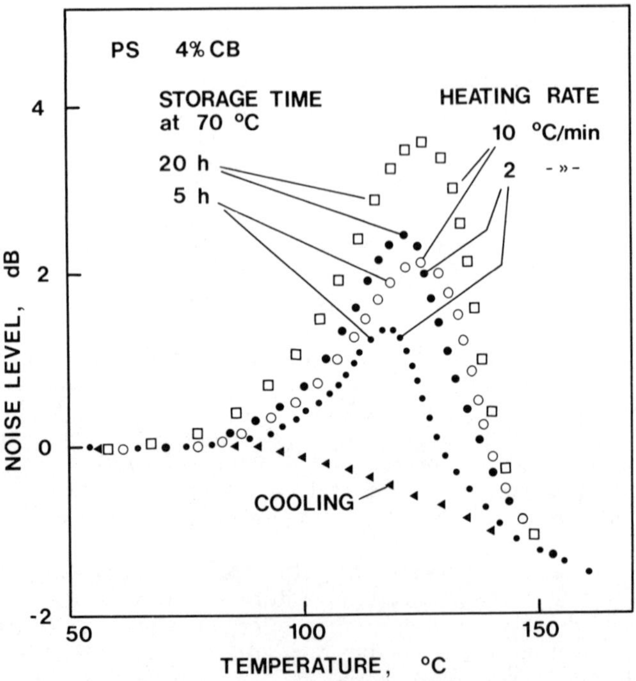

Figure 1. Effect of storage time and heating rate on thermal noise in the T_g *region. PS, 4% carbon black (CB). Bandwidth, 500–1500 Hz; thermal noise level, 0 dB at 25°C.*

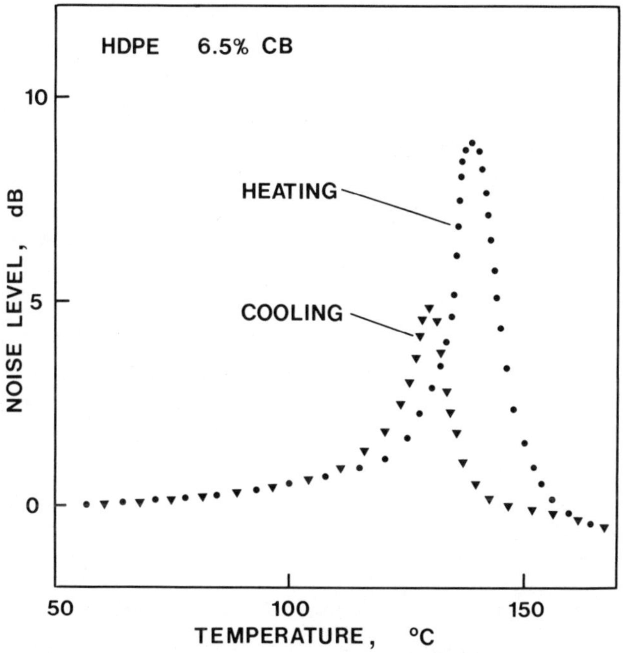

Figure 2. Thermal noise vs. temperature for HDPE containing 6.5% CB during heating and cooling around T_m *region. Heating rate, 1°C/min; bandwidth as in Figure 1.*

The measuring cell was made of PTFE and had a four-electrode system connected to its outer cylinder. One pair of electrodes served as terminals for the measuring amplifier, and the outer pair was connected to the current supply through a 10-MΩ wire-wound resistor (the sample resistance was much less than 10 MΩ). The measuring cell was carefully shielded from electric and magnetic fields and from mechanical vibrations.

Measuring Equipment. The noise measuring amplifier (PAR model 113) was a low-noise device with a very high input resistance. The noise figure of the amplifier had to be low (less than 3 dB) for the actual resistance of the sample and the frequency range determined by the capacitance over the input terminals. The signal from the noise amplifier was fed to a band pass filter adjusted to the useful range of frequencies to optimize the signal-to-noise ratio. It was necessary to have several different amplifiers to match different sample resistances. The signal was finally recorded and analyzed using a Brüel and Kjaer frequency analyzer (model 2120).

The sample resistance (ac) calculated from the observed thermal noise voltage spectra using the Nyquist formula was compared with the values measured with a conventional resistance bridge (General Radio, model 1620). Similar measuring equipment has been described in previous publications (*3, 4*).

Results

Noise in Carbon-Black-Filled PS and HDPE around T_g and T_m. Because of high resistivity values, it has not been possible to measure the noise characteristics of pure polymer samples. To reduce the resistivity to an acceptable level, carbon black (4–6.5%) was added to the polymer under investigation. The resistivity level of the samples was 10^4–10^6 ohm-cm.

The general features of the noise behavior around T_g and T_m are illustrated in Figures 1 and 2, showing the temperature variation of the thermal noise level over the frequency range 500–1500 Hz for PS at T_g and HDPE at T_m. In both cases distinct peaks were found in the T_g and T_m regions, respectively. The T_g peak exhibited a pronounced time dependence as illustrated in Figure 1. For instance, no peaks were observed when cooling the sample from $T > T_g$. Neither was there any peak when such a sample was immediately reheated; storage at $T < T_g$ was necessary for the peak to redevelop to full intensity. The necessary

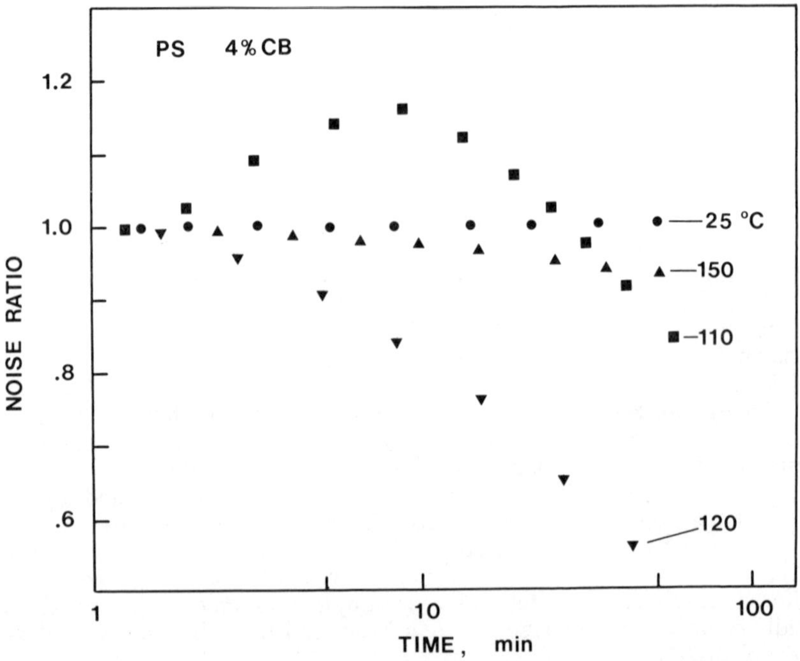

Figure 3. Thermal noise level ratio (noise time t/noise at 1 min after reaching temperature) vs. time at temperatures far from (25°, 150°C) and close to (110°, 120°C) temperature at which noise peak occurs (120°C) for PS stored 10 hr at 70°C. Samples heated to indicate temperatures at 20°C/min. Bandwidth, 500–1500 Hz.

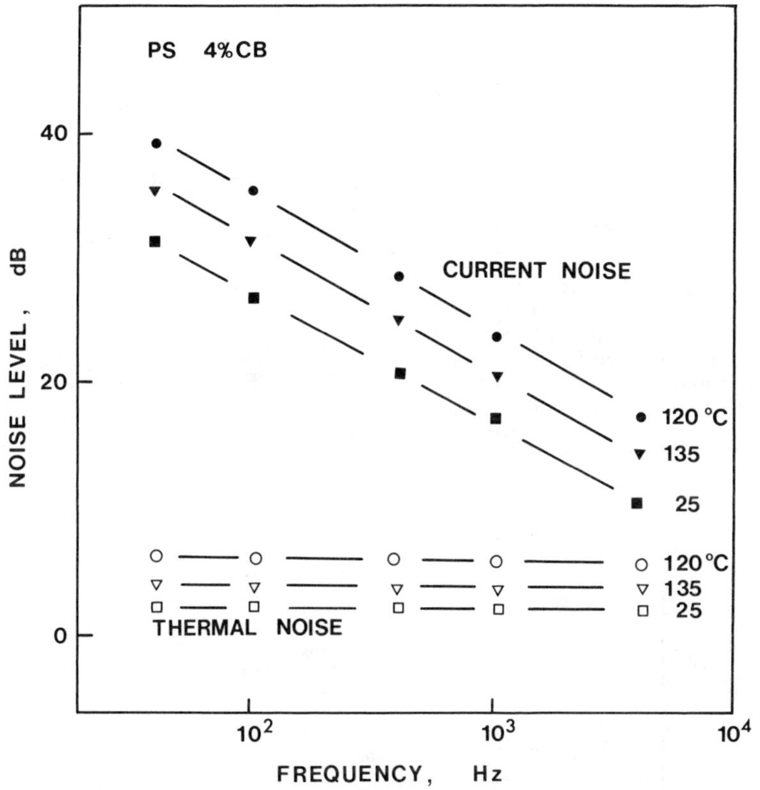

Figure 4. Noise spectra at various temperatures for a PS sample (4% CB) showing an 1/f type dependence of the current noise (direct current 20 μA) and a frequency independence of the thermal noise at all temperatures (even in the transition region). Bandwidth, 2 Hz.

storage time was typically about 5 hr (at 70°C); it increased with decreasing storage temperature. The time dependence of the T_m peak of HDPE was less pronounced than that of the peak at T_g (PS). Again, this dependence was confined to the transition (melting) region only. An example of the effects observed is the heating/cooling hysteresis in peak shape shown in Figure 2, which is reminiscent of similar effects encountered, for instance, in measurements of various thermal parameters around T_m (5).

Further details of the time dependence of the T_g peaks as observed with carbon-black-filled PS are evident from Figure 3, which shows the variation of the noise intensity (relative to its 1-min value) with time at different temperatures. The time dependence of this intensity is most pronounced in the vicinity of the noise peak temperature, although it

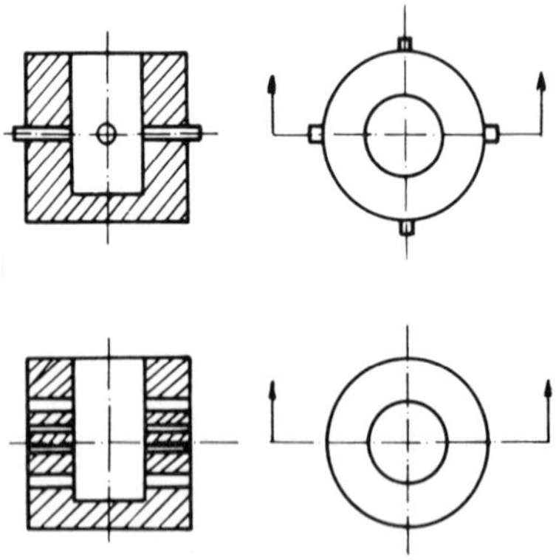

Journal of Applied Physics

Figure 5. Diagram of platinum electrode arrangement in the outer cylinder of the measuring cell. (top) *Four Pt rods; diameter, 6 mm;* (bottom) *four Pt rings; thickness, 2 mm. Diameter of rotating bob* (not shown) *24 mm; height, 50 mm. Gap between bob and cylinder, 1.5 mm (4).*

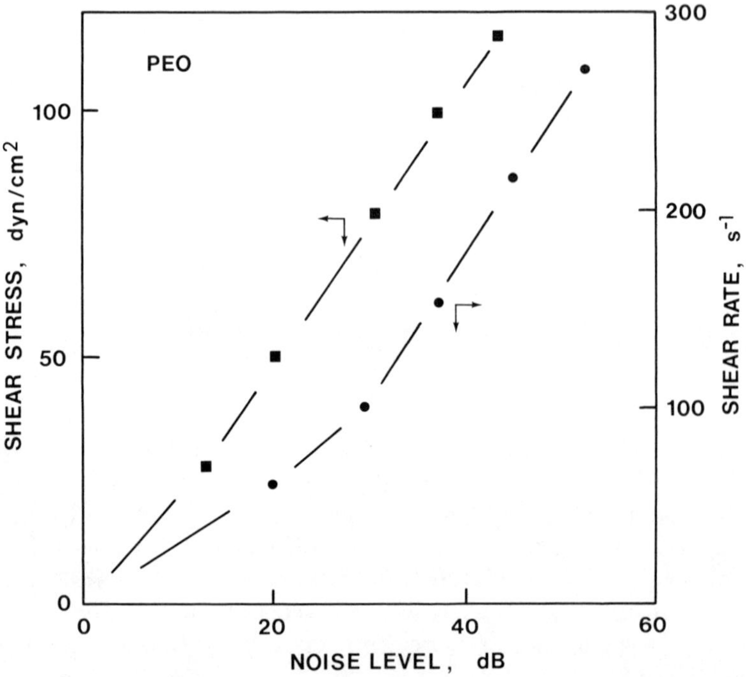

Figure 6. Current noise level vs. shear stress/rate at 10 Hz for aqueous solution of PEO (M_w *4 · 10*^{6}*, 0.5% PEO). Direct current, 20 μA, noise level at zero shear, 0 dB.*

remains practically constant outside this region (25°C, 150°C). The noise peaks did not occur exactly at the T_g value of the polymer under study but 10°–30°C higher. For HDPE, the difference between the noise peak position and T_m was comparatively small (3°–5°C).

The frequency distribution of the current noise was of the $1/f^{\alpha}$-type and was independent of temperature; that is to say, it was the same inside and outside the T_g or T_m region, respectively. This constancy is illustrated in Figure 4 for the T_g region of PS. The spectrum for thermal noise was white in all the measurements carried out. The sample resistance values calculated from the observed noise spectra agreed with the resistance values obtained with the conventional resistance bridge irrespective of the temperature or time scale of the experiment.

Besides being substantially higher, the current noise level varied with temperature in about the same way as the thermal noise varied in the T_g and T_m regions of the two polymers investigated. Again, the peaks appeared at temperatures somewhat higher than those corresponding to

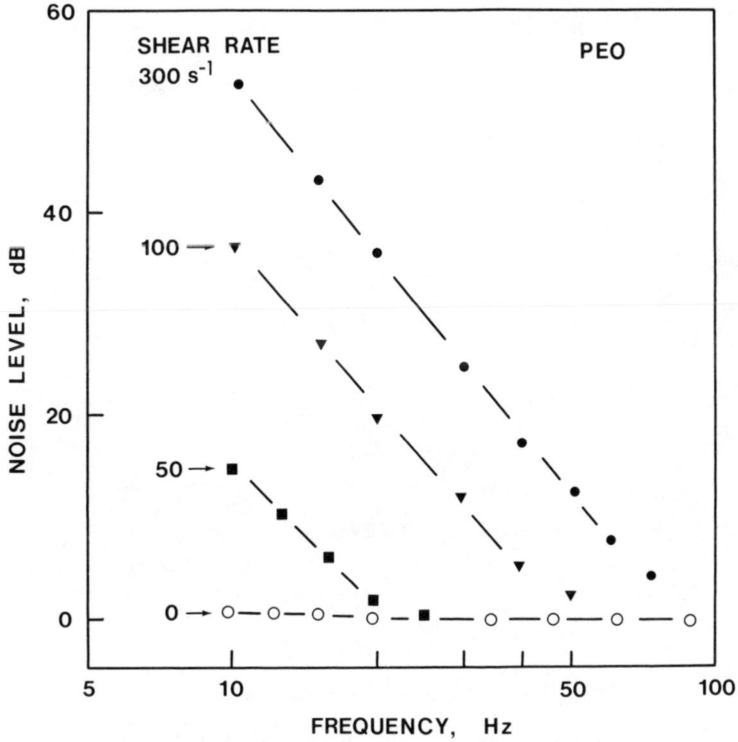

Figure 7. Current noise spectra for different shear rates. Sample and conditions as in Figure 6.

the normal T_g or T_m positions obtained using a similar time scale. Experiments with several other amorphous and crystalline polymers (PMMA, PVC, PVAc, LDPE, and PP) gave similar results.

Noise in Aqueous Solutions during Couette Flow. Using the cells shown in Figure 5, thermal and current noise levels were determined for aqueous solutions of PEO of varying molecular weight. The significance of the results was corroborated by measurements on aqueous solutions of other polymers.

With regard to thermal noise, only weak indications of an increased noise level when increasing the shear rate were found. These effects will thus not be discussed here. When recording the current noise, however, the noise level increased significantly with the shear rate (stress). Figure 6 provides a typical illustration (PEO solution in water).

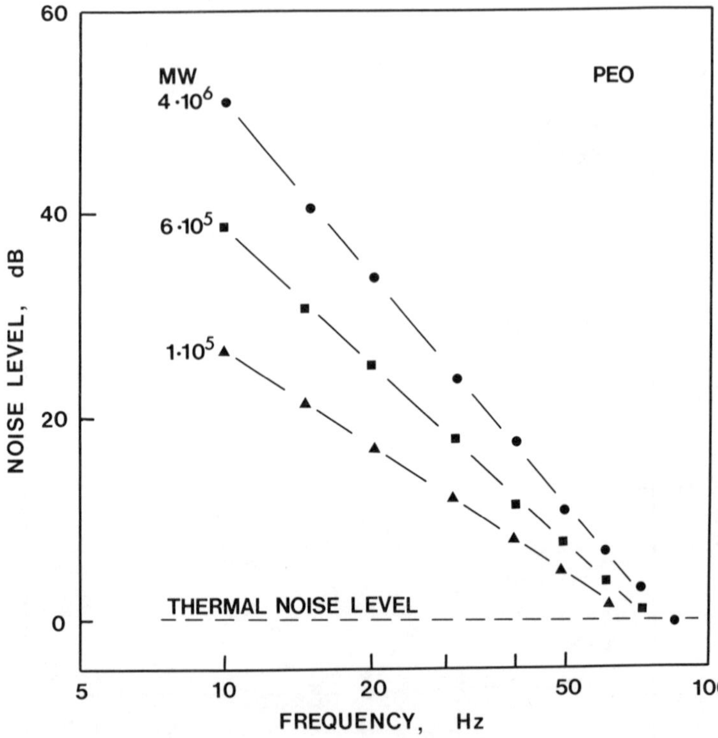

Figure 8. Effect of molecular weight on current noise spectra for aqueous solutions of PEO having approximately equal apparent viscosity at a shear rate of 250 sec^{-1}. (top) 1% PEO, $\overline{M}_w$ 4 · 10^4; (middle) 4.3%, $\overline{M}_w$ 6 · 10^5; (bottom) 6%, $\overline{M}_w$ 10^5. Direct current, 20 μA; thermal noise level, 0 dB.

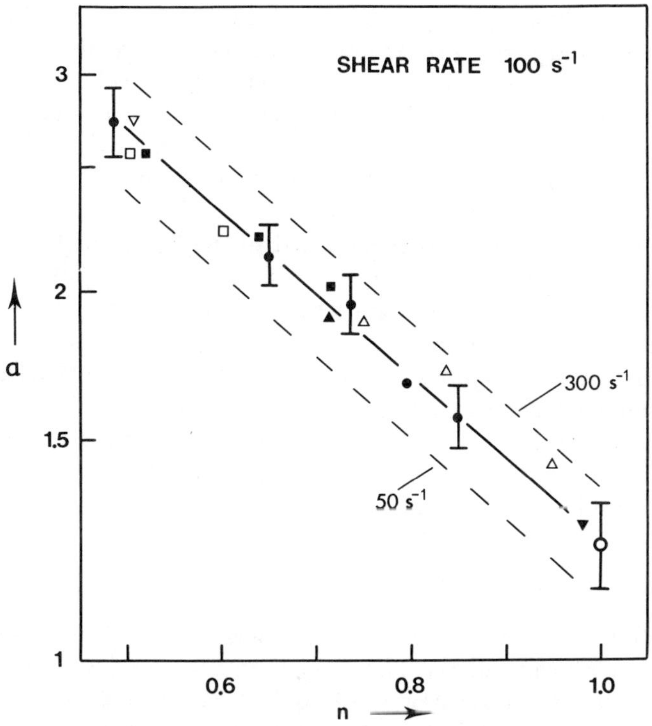

Figure 9. Plot of α vs n for aqueous solutions of different polymers. (●) PEO 0.2, 0.3, 0.5, and 1.0% (w/w) type WSR301; (▲) PEO 4.3% WSR205; (▼) PEO 10% WSR-N10; (▽) HEC 0.2%, QP100M, (□) HEC 0.2% and 0.5% QP-52000; (■) CMC 0.2%, 0.7%, and 1%; (△) PVP 5.7% and 10%; (○) PVA 5%. Upper and lower curves represent α(n) relation at 300 and 50 sec^{-1}, respectively.

The frequency distribution of the current noise was of the $1/f^{\alpha}$ type. Figure 7 shows this together with the influence of the shear rate. The increase of the slope of the noise level/frequency lines (= exponent α) with the shear rate was relatively limited. Much more pronounced was the influence of increasing the molecular weight of PEO as shown in Figure 8. Changes in the molecular weight also produced a varying degree of pseudoplasticity as expressed by the exponent n in the Ostwald–de Waele relation

$$\tau = KD^{n} \tag{2}$$

where τ and D denote the shear stress and shear rate, respectively, and K the consistency. It is therefore rather natural to compare the exponent α in the $1/f^{\alpha}$ frequency distribution with the exponent n of Equation 2.

The result is shown in Figure 9, reproducing the $\alpha(n)$ relationship for PEO solutions at varying shear rates and molecular weights and also for several other aqueous polymer solutions. The experimental points arrange themselves comparatively well along the line log $\alpha(n)$. However, the character of this relation is approximate. For instance, the line shown in Figure 9 will undergo a slight vertical shift when the shear rate is changed.

The extent of this shift is indicated in the figure for the two shear rate extremes of 50 sec^{-1} and 300 sec^{-1}, respectively. This shift is comparatively small and does not invalidate the general character of the relationship referred to in Figure 9. Apart from PEO, the data of this figure relate to aqueous solutions of hydroxyethyl cellulose (HEC), sodium carboxymethyl cellulose (CMC), polyvinylpyrrolidone (PVP), and poly(vinyl alcohol) (PVA). The behavior of the solutions at shear rates lower than 50 sec^{-1} was not measured because of the low intensity of the shear-induced current noise.

Discussion

Within the range of frequency, temperature, shear rate etc. covered by the experiment, all the measured thermal noise levels agreed well with the predictions based on the Nyquist formula. This implies that the thermal noise level could have been calculated from resistivity measurements and also that the noise peaks in the vicinity of T_g and T_m would have appeared in the corresponding resistivity–temperature diagrams. This was actually verified in numerous experimental runs. On the other hand, the measurement of thermal noise has the advantage that no external voltage has to be applied across the sample. This eliminates the possibility that the observed peaks arise from polarization effects (*6,7*).

It is still not clear whether the measurements reported above are an inherent property of the polymer or whether they are associated with the presence of carbon black. Probably the latter explanation is appropriate since the disruption and reformation of a carbon particle network at the critical temperatures is a likely interpretation of the effects observed (*1*). In this respect, the noise peaks should be distinguished from resistivity peaks measured at T_g on pure polymers. The latter have been shown to be caused by polarization effects (*6*).

The temperatures at which the noise peaks occur are higher than the corresponding T_g and T_m peaks observed with other methods such as DSC or TSC. Further, the effects of storage time and heating rate are different. This indicates that the noise method senses other structural changes than those associated with the DCS or TSC peaks. The noise peaks appear to be caused by a complex interaction of the carbon particle network and the polymer. No noise peaks are observed in the T_m region of low-molecular-weight olefins and similar substances (*13*).

The fact that carbon-black-filled polymers had to be used in this work is a rather unfortunate complication with regard to the immediate applicability of the results. In this respect, the current noise in polymer solutions during shear can be interpreted more readily since the recorded effects do not appear to be accompanied by corresponding changes in resistivity. On the other hand, the interpretation of the results is complicated by the fact that there is no accepted theory of the $1/f$ noise despite numerous recent attempts to produce a generally valid explanation of such fluctuations (*8, 9, 10, 11, 12*).

One of the limitations of this study was the lower frequency limit of ca. 10 Hz. Our recent measurements of noise effects associated with the flow of polymer solutions at frequencies down to 5 mHz have shown that this low frequency region is especially interesting. For instance, thermal noise increased with the shear rate, exhibiting a $1/f^{\alpha}$ frequency distribution for the solutions discussed.

Literature Cited

1. Klason, C., Kubát, J., *J. Appl. Polym. Sci.* (1976) **20**, 489.
2. Nyquist, H., *Phys. Rev.* (1928) **32**, 110.
3. Klason, C., Kubát, J., *J. Appl. Polym. Sci.* (1975) **19**, 831.
4. Klason, C., Kubát, J., *J. Appl. Phys.* (1976) **47**, 1970.
5. Ke, B., *Polym. Rev.* (1964) **No. 6.**
6. Seanor, D. A., *Tech. Methods Polym. Eval.* (1970) **2**, 293.
7. Carley Read, R. E., Stow, C. D., *Br. J. Appl. Phys.* (1969) **2**, 567.
8. Weissman, M. B., *J. Appl. Phys.* (1977) **48**, 1705.
9. Tandon, J. L., Bilger, H. R., *J. Appl. Phys.* (1976) **47**, 1697.
10. Hooge, F. N., *Physica* (1976) **83B**, 14.
11. Bittel, H., *Physica* (1976) **83B**, 6.
12. Stephany, J. F., *J. Appl. Phys.* (1975) **46**, 665.
13. Bueche, F., *J. Polym. Sci.* A-2 (1973) **11**, 1319.

RECEIVED November 10, 1977.

2

Acoustic Emission of Polymers under Tensile Load

A. PETERLIN

Institute for Materials Research, Polymers Division, National Bureau of Standards, Washington, DC 20234

The observation of acoustic emission from a crazing polymer depends very much on the plastic deformability of the polymer. The crazes in ductile materials propagate so smoothly by relatively slow plastic deformation at the tip of the craze that the acoustical signal is not strong enough to be unambiguously separated from the high noise level. It is only in very brittle material that the jumpwise craze propagation yields acoustic bursts strong enough to be recorded easily. Such materials are polyvinyltoluene with a strain-to-break of less than 1%, low molecular weight polystyrene, and aged poly(methylmethacrylate). For brittle composite material it is better if the components differ greatly in their mechanical properties.

Acoustic emission during metal deformation was first systematically investigated by Kaiser (*1*) who used electronic instrumentation to listen to the sound emitted by zinc, steel, aluminum, copper, and lead samples during tensile loading. He attributed the acoustic emission to grain boundary sliding. He also observed that no acoustic emission was generated during second loading until the stress exceeded the maximum value of the first run. This effect has become known as the Kaiser effect and has proven to be very useful in acoustic emission studies of the mechanical history of metals. Since practically no acoustic emission takes place up to the maximum load the sample has experienced at any tensile straining after its last annealing, one has a simple method to detect this maximum exposure without destroying the sample.

Later the instrumentation was improved and the sources of acoustic emission investigated more thoroughly. Excessive background noise was reduced drastically by working with instrumentation systems whose frequency range was well above the audio range (*2*). As a rule, one presently works in the range between a few hundred and a few thousand kH. It was also found that the emission from strained metals is mainly caused by the motion of dislocations in crystal lattice (*3*) and by phase transitions (*4*) during plastic deformation. The actual plastic deformation which causes acoustic emission occurs mainly at crack tips and other stress raisers when a flawed specimen is loaded. Hence, the technique can be used to detect flaws, to study crack growth, and to monitor the structural integrity of samples. Triangulation techniques similar to those used to locate earthquakes can help to locate flaws in a structure under study. It is an important nondestructive testing technique widely accepted in work with metals.

Acoustic Emission in Polymers

Sound emission was noticed in polymers by Sauer and Hsiao (*5*) during experiments on crazing of polystyrene. They stated that "usually, sharp cracking sounds could be heard soon after the load was applied which suggests the sudden appearance of crazing and the immediate release of a portion of energy into sound waves" but did not pursue the effect any more. Grabec and Peterlin (*6*) were the first to apply the technique of acoustic emission to polymers to study the effect as a function of deformation, crazing, and mechanical history. They used an extremely brittle glass, polyvinyltoluene, which at room temperature breaks at a strain of 0.45%, i.e., still in the perfectly linear stress–strain range.

They observed a slightly more than linear increase of acoustic activity with increasing strain and stress, with the close correlation of acoustic bursts with craze initiation and growth, and the validity of the Kaiser effect. Practically no acoustic emission is detectable during straining up to the point where the first crazes appear. The opening of crazes seems to be sufficiently rapid to produce an acoustic burst. The same applies to stepwise craze propagation. In both cases enough energy must be accumulated which is suddenly released when the craze opens or expands. If one removes the load and repeats the straining, one does not record any acoustic emission up to the maximum load or strain of the previous run. The emission shows up again as soon as one surpasses this level. Moreover, the emission does not resume full strength immediately so that the new emission intensity is below the intensity expected from continuation of the emission intensity of the first run. But, soon the emission increases a little faster thus compensating for the initial defect. After this compensation seems to be completed, one is again back at the intensity obtained in a single uninterrupted run extending up to sample failure.

Kramer and co-workers (7) reported that acoustic emission occurred in polystyrene immersed in different swelling liquids only when the crazes ruptured but not during their formation and growth. As long as the bridging by filamental elements is still intact, the deformation at the craze tip and in the craze is still so slow that no acoustic bursts are generated. It is the final fracture of these elements which is abrupt enough to cause the emission of a detectable acoustic burst.

Recently at NBS we observed acoustic emission of strained poly (methyl methacrylate) (8). In order to reduce the noise background from a tensile testing machine, the samples were investigated in a bending mode. Crazing was enhanced by some acetone dropped on the bent sample. The formation of crazes and the acoustic emission were recorded on magnetic tape so that the effects observed could be investigated thoroughly under repeated viewing. We had two different samples, one kept at room temperature and one exposed for a long time to substantially higher temperatures. The crazes turned out to be quite different in both samples. In the first material the crazes were very fine and grew smoothly through the whole sample exactly perpendicular to the local stress. Little if any coalescence of close-by crazes was observed. The second sample has shown many small crazes which grew mainly by coalescence with close-by crazes even if the connecting section was not perpendicular to the local stress. The growth seemed to be extremely discontinuous. For a long while the crazes did not grow appreciably, and then suddenly they extended to the next craze. Acoustic emission was observable only in the second material, but in the first one the bursts, if there were any, did not surpass the noise background.

The explanation of this aging effect can be based on the common concepts of aging. The initial material is ductile and tough enough that the plastic deformation under tensile load proceeds relatively slowly, yielding weak acoustical signals which do not surpass the noise background.

Aging at higher temperatures may change the fractional free volume, let evaporate some plasticizer, reduce the molecular weight, and add some oxygen-containing end groups. The modification of the sample seems to occur rather irregularly, i.e., some areas are substantially more damaged than the rest. In such a material the crazes originate at places with maximum damage concentration but cannot extend into the much-less-damaged surrounding areas before a substantial increase in stress level which provides enough energy accumulation for such an expansion. Hence, the growth of crazes is mainly achieved by sudden jumps. Since the energy available is so large, the jumps can deviate substantially from perpendicularity to stress and thus reach close-by crazes and coalesce with them. The generation of isolated crazes and the jumps seem to be so abrupt that a substantial fraction yields recordable acoustic emission.

Sauer (9) recently stated that the polystyrene investigated by him and Hsiao (5) was a low-molecular-weight material and hence was more brittle than the presently produced material which has a higher average molecular weight and consequently is more ductile. This may be the best explanation why acoustic emission in polystyrene has not been subsequently reported by any worker investigating crazing effects.

On the basis of all these experiments on acoustic emission in polymers one can conclude that the observability of the effect depends very much on plastic deformability of polymers. Crazes in ductile material propagate so smoothly by relatively slow plastic deformation at the tip of the craze that the acoustical signal is not strong enough to be separated unambiguously from the high noise level. It is only in very brittle material that the jump-wise craze propagation yields acoustic bursts that are strong enough to be recorded easily. Such materials are polyvinyltoluene with a strain-to-break of less than 1%, low-molecular-weight polystyrene, and aged poly(methyl methacrylate). In the last case the fresh polymer does not yield any detectable acoustic emission. The crazes grow smoothly through the whole sample and are exactly perpendicular to the local stress. In aged material with highly localized damage concentration the generation of isolated crazes and the abrupt jumps from one craze to the next one are sufficiently rapid effects for such intense acoustic emission that single bursts can be recorded.

Microcrack Opening as the Source of Acoustic Emission

The time dependence of local displacement ϵ and displacement velocity $d\epsilon/dt$ at the craze or crack initiation is shown schematically in Figure 1. The velocity drops to one half of its maximum value at $t_{1/2}$. If one puts $t = 0$ at the maximum and assumes a symmetric time dependence of displacement velocity, the half width of velocity distribution at such an elementary act is $2t_{1/2}$. The Fourier transform (FT) of velocity yields the spectral distribution of the emitted acoustic burst. The intensity is a maximum at zero frequency and drops to half this value at $\omega_{1/2} = 2\pi\nu_{1/2}$. In the first approximation the product of the half width of velocity and frequency distribution of acoustic emission, $2t_{1/2} \times 2\nu_{1/2}$, equals 0.8825. To have a substantial amount of energy available in the frequency range of 1 MHz ($= \nu_{1/2}$) the displacement velocity curve vs. time must have a half width of 0.5 μsec, i.e., the major part of the local displacement must occur within 0.5 μsec. The square of the maximum value of velocity times

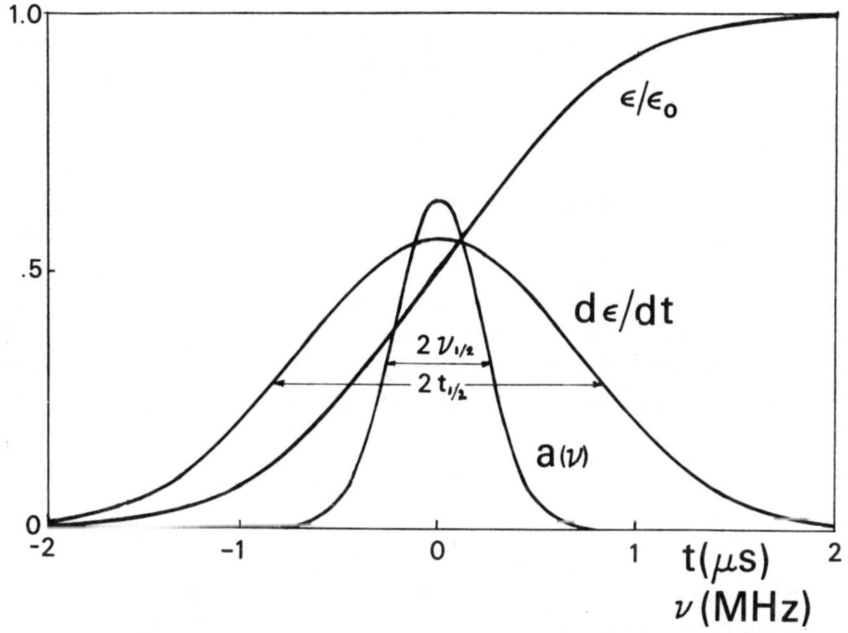

Figure 1. Schematic of a symmetric deformation ε, deformation velocity dε/dt at the initiation, and formation of a craze or crack as functions of time and the Fourier transform of velocity distribution with $\nu_{1/2} = \ln 2/\pi t_{1/2} = 0.2206/t_{1/2}$

the mass of the volume affected yields the energy content of a single burst of acoustic emission connected with the craze or crack initiation. For the same displacement the maximum velocity, and hence also the total amplitude of acoustic emission, is increasing as the inverse half time of deformation velocity. Since the total energy in an acoustic burst increases as the square of and the acoustic frequency range extends as the inverse half time of the deformation rate, one sees that the chances for so intense acoustic bursts that they can be separated safely from the noise background increases extremely rapidly with the abruptness of deformation. Any ductility of the material involved increases this time and thus reduces the chances to observe acoustic emission.

Since the finite ductility of polymer glasses is the main obstacle against sufficiently strong acoustic emission which can be easily separated from the noise background, one is tempted to reduce it by choosing either a material with a very high T_g or by lowering the temperature of the experiment. One has little choice in the former case and has plenty of additional noise in the latter case. We have not yet successfully performed any really satisfactory experiment along these lines.

One reason for this absence of easily detectable acoustic emission from uniform polymer glasses seems to be the fact that even far below the T_g the ductility at the crack tip is many times higher than that of the bulk sample. Hence it is too high for the sufficiently abrupt deformation needed for a strong acoustic emission during craze or crack initiation. The conclusion is corroborated by the well known fact that in most polymers investigated even 100°–200°C below the T_g temperature, the surface of the brittle crack shows a well developed layer which was formed by local ductile deformation in front of the propagating crack (*10*). The bulk sample is brittle as far as the location of the breaking point on the stress–strain curve is concerned. In the microscopic region in front of the crack tip, however, the ductility is sufficient for small-scale plastic deformation yielding the crazed surface layer of the crack.

At liquid nitrogen temperature such an effect may be caused by the plasticizing action of air or nitrogen as amply demonstrated by crazing experiments on amorphous (*11*) and semicrystalline (*12*) polymers in contact with air, nitrogen, oxygen, and argon. The effect is repeated with carbon dioxide at − 80°C. In vacuum or in helium the plasticizing effect does not occur. As a consequence, in this environment the sample breaks in the most brittle manner (*13*). Hence experiments on acoustic emission at low temperature will have to be performed in helium to avoid the plasticizing effect of the gases at high activity, $a \sim 1$, which may increase the ductility so much that the acoustic emission is not sufficiently strong or at least not concentrated at sufficiently high frequencies for easy detection.

Acoustic Emission in Polymer Composites

Better candidates for the study of acoustic emission are composite materials. The extremely brittle polyvinyltoluene sample which showed easily detectable acousic emission (*6*) was indeed to some extent such a composite material since it was a sample used for scintillation counting of nuclear radiation. The crystalline particles of the inorganic scintillator embedded in the rather rigid polymer matrix differ enough in elastic properties from those of the matrix that a substantial stress enhancement occurs on the interface between the two components. One has about twice the bulk stress on the poles and one third on the equator of a perfectly rigid spherical particle. Such a stress increase in the poles leads rather early to adhesion failure of the particle–matrix boundary and to microcrack formation. This finally makes the sample fail at small strain-to-fracture, $\epsilon_b = 0.5\%$. The microcracks act as nuclei for crazing. The opening of a fissure between the particle and the matrix is sufficiently abrupt for a detectable acoustic emission which is closely correlated with craze initiation or even growth.

Such an explanation was corroborated by uniaxial tensile experiments on composite samples containing 10–40% small glass beads of about 15 μm in diameter embedded in a hard epoxy matrix (*14*). The observed acoustic emission was roughly proportional to the glass content, i.e., to the number of glass beads per unit volume. Hence one concludes that the origin of the recorded emission was in the boundary between the very hard glass sphere and the epoxy resin matrix where the failure of adhesive bonding produced microcracks.

A close inspection of the strained sample has shown a great many small cracks at the sharp teeth of the grips which penetrated the surface of the sample. The indentations grew with increasing elongation, i.e. with increasing load. The growth of these small cracks seems to be the source of the acoustic emission. Practically no crazes could be observed on the rest of the strained sample.

The acoustic emission from such uniaxially strained composite material occurs in rather strong bursts which greatly surpass the noise background. If the load is removed and the sample left to relax, one does not record any emission upon new loading up to the maximum strain of the first loading. The experiment can be repeated by interrupting the loading at different levels. New emission only starts after the maximum strain of any previous loading is reached and surpassed. The ensuing emission takes place at the same slope as was observed before the interruption of loading.

The load–elongation curve of the second run is below that of the first run up to the maximum strain of the first straining and is a smooth continuation of that of the first run as soon as the strain goes to higher values. This confirms the basic assumption and explanation of the Kaiser effect in polymers that during the second loading one only reopens the microcracks formed during the first run but does not form any new one. The reopening is a much less disruptive process than the first formation of a microcrack. Hence it does not require the same stress and also does not yield any acoustic emission.

Acoustic Emission in Fibrous Polymers

This experience so clearly demonstrated on the brittle composite materials can be applied immediately to the deformation of fully drawn fibrous material which may be extended up to fracture. If one stops a little before rupture and removes the load completely, the sample will slowly approach almost zero strain. The deformation is nearly completely recoverable. The next loading yields a lower load–elongation curve. One has tried to explain the difference as a consequence of fractured chains which under favorable conditions can be detected by the electron spin resonance (ESR) of free radicals formed by the new chain ends (*15*). It turned out that this number is too small for such an explanation by a few orders of magnitude.

During loading one creates plenty of microcracks which can be detected by small angle x-ray scattering (*16*). During unloading the microcracks close again since the sample strain goes practically to zero. In the second run the reopening of the same microcracks is much easier than their initial formation. In the first run one had to act against molecular cohesion of the original material and eventually had to break all tie molecules bridging the microcrack. Under favorable conditions the fractured molecules with end-of-chain radicals or some secondary middle-of-chain radicals are detectable by ESR. In the second run one reopens the microcracks. This time one has to overcome only a small remnant of molecular cohesion since closing of the crack during unloading is never perfect on a molecular scale, and it is not necessary to rupture any tie molecules since all were ruptured during the first opening of the microcracks. This difference in work requirement and stress is reflected in the lower load–elongation curve of the second run.

In fibrous material the candidates for microcrack formation are the ends of microfibrils where the normal axial connection of folded chain crystal blocks by taut tie molecules is either completely missing or so much reduced that the resulting local elastic modulus is much less than in the interior of the microfibril where the amorphous layers are bridged by so many taut tie molecules (*17*). At such point defects of the microfibrillar lattice the axial transmission of tensile stresses is less efficient and yields larger strains than in the rest of the sample. As a consequence of the relatively easy axial sliding motion of short sections at the end of the microfibril, the strain in such a defect area may soon be so large that it leads to microcrack formation. The few taut tie molecules in the area are overstrained so much during this process that they break. The ruptured chains can be detected by ESR if the primary or secondary radicals are stable enough.

Such a microcrack formation occurs as soon as the strain enhancement at any microfibril end is high enough for material separation. The higher the bulk strain the higher the number of such defects which are deformed so much that a microcrack can be opened. Their number increases almost exponentially with strain as can be concluded from the dependence of radical concentration on bulk strain. But the sample itself is still strong and will hold the load up to the point where the microcrack coalescence yields the first critical size crack which will start to grow catastrophically and will make the sample fail.

The maximum energy released during the primary opening of such a microcrack with average dimensions about 100 Å (average diameter of microfibril) is about $\beta V E \epsilon_b^2/3 \sim 10^{-8}$ erg if one takes for the chain $E = 240\ GP_a$, $\epsilon_b = .3$, $V = 10^6 Å^3$, $\beta = .1$ (10% of chains in amorphous layers are taut tie molecules). It seems unlikely that the acoustic emission associated with such an elementary act could be detected. The chances are better with microcrack coalescence along the outer boundary of fibrils which involves a rapid sequence of microcracks opening and eventual full separation of the fibril from a fraction of adjacent fibrils. On the other hand one can expect that with increasing strain the frequency of microcrack formation will be so high that the cumulative acoustic emission from a finite volume of the sample will be detectable above the noise background.

Literature Cited

1. Kaiser, J., *Arch. Eisenhuettenwes.* (1953) **24,** 43.
2. Dunegan, H. L., Tatro, C. A., Harris, D. O., "Acoustic Emission Research," *Lawrence Livermore Lab (Rep)* (1964) **UCID-4868,** Rev. 1.
3. Schofield, B. H., "Acoustic Emission under Applied Stress," Report **ARL-150,** Aeronaut. Res. Lab., Off. Tech. Serv., Washington, DC, 1961.
4. Liptai, R. G., Dunegan, H. L., Tatro, C. A., *Int. J. Nondestr. Test.* (1969) **1,** 213.
5. Sauer, J. A., Hsiao, C. C., *Trans. ASME* (1953) **75,** 895.
6. Grabec, I., Peterlin, A., *J. Polym. Sci., Polym. Phys. Ed.* (1976) **14,** 651.
7. Kramer, J., personal communication.
8. Green, R., Penn, R., Peterlin, A., unpublished work.
9. Sauer, J. A., personal communication.
10. Kambour, R. P., *Macromol. Rev.* (1973) **7,** 1.
11. Parrish, M., Brown, N., *Nature (London), Phys. Sci.* (1972) **237,** 122.
12. Olf, H., Peterlin, A., *Polymer* (1973) **14,** 78.
13. Olf, H., Peterlin, A., *J. Polym. Sci., Polym. Phys. Ed.* (1974) **12,** 2209.
14. Green, R., Penn, R., Peterlin, A., Smith, J., unpublished work.
15. Lloyd, B. A., DeVries, K. L., Williams, M. L., *Rheol. Acta* (1974) **13,** 352.
16. Zhurkov, S. N., Slutsker, A. I., Kuksenko, S., *Fiz. Tverd. Tela* (1969) **11,** 190.
17. Peterlin, A., *J. Macromol. Sci.* (1973) **B7,** 705.

RECEIVED November 10, 1977.

3

The Nanotensilometer—An Accurate, Sensitive Tensile Test Instrument

C. G. ANDEEN and R. W. HOFFMAN[1]

Department of Physics, Case Western Reserve University, Cleveland, OH 44106

A new instrument has been constructed to measure force-elongation data of thin films and fibers with cross sectional areas as small as 10^{-14} m^2. The nanotensilometer can measure forces of 1×10^{-8}–5×10^{-1}N. Elongations of 0.1–10^6 nm are measured to an accuracy of 0.1% by a differential three-terminal capacitor. The instrument provides direct plotting of force–elongation curves. For solution-grown polyethylene single crystals about 10 nm thick and 15 μm in planar dimensions, the force rises almost linearly to a maximum at about 0.5 dyn corresponding to a stretching of 50 nm from an initial gage length of 2 μm. Appreciable forces are still observed for strains of 500%.

A new unique instrument, which we have named the nanotensilometer, has been designed and built for the express purpose of determining the tensile properties of polymer single crystals and other such samples with exceedingly small cross sectional areas. Several new concepts were incorporated in the design of the nanotensilometer to obtain the accuracy and sensitivity needed to handle samples having cross sectional areas as small as 10^{-14} m^2. The constraints of one-dimensional motion and extreme isolation from ambient vibration were met by constructing two statically and dynamically matched platforms which move relative to each other. This chapter describes the concepts that resulted in the nanotensilometer design, describes the instrument and its operating specifications, and presents preliminary data for polyethylene single crystals.

[1] To whom correspondence should be addressed.

0-8412-0406-3/79/33-174-025$05.00/1

Previous tensilometers for thin-film samples are discussed in a review by Hoffman (*1*). They represent extensions to higher sensitivities of conventional tensile test devices. They may exist as a "hard" or "soft" instrument depending on whether a constant strain rate or a force is applied and whether a force or elongation is measured. The specimens for use in such instruments were thin in one dimension but had other dimensions of at least a millimeter. Great care needed to be taken with the design and the fastening of the thin-film sample to the jaws of the instrument. Often the tensile data and especially the conditions at failure of the specimen depended on the difficulties of mounting. The results of such studies indicated that the elastic moduli were those expected from measurements on bulk samples except for lower elastic moduli observed for amorphous metallic films compared with the crystalline phase. Fracture generally occurred at strains of a few percent. Plastic deformation was observed at relatively low strains with approximately three fourths of the deformation at fracture being elastic.

Metallic whiskers have long been studied because of their high strengths resulting from almost dislocation-free structures. They have been examined in tension by Brenner (*2*). More recently Riley and Skove (*3*) have shown that non-Hookian behavior may result from the effect of higher order elastic constants in single-crystal whiskers. They also developed a tensilometer using a LVDT sensor capable of measurements with metallic whisker samples. Natural and synthetic fibers have also been studied. The most common instrument today for tensile properties is an Instron for which specimens commonly have cross-sectional areas $> 10^{-9}$ m^2 and gage lengths > 0.01 m. Typical strain rates are 100% per min. Stress relaxation data have also been reported by observing the decrease in force as a function of time at a constant elongation. Dynamic measurements of the loss peaks are also a common mechanical measurement (*4*).

Biological fibers have been investigated (*5*). However, in all the tensile testers referenced above, at least one dimension of the sample has been larger than a millimeter. Such devices are not useful for examining polymer single crystals in which the sample size consists of small platelets with in-plane dimensions of 10 μm and a thickness of some 10 nm.

Nanotensilometer

A brief description of the nanotensilometer is published elsewhere (*6*), and a detailed discussion is in preparation. Thus we consider here only the most important features of the design and some of the operational characteristics. The key feature in the design of this new instrument is the construction of two statically and dynamically matched inverted

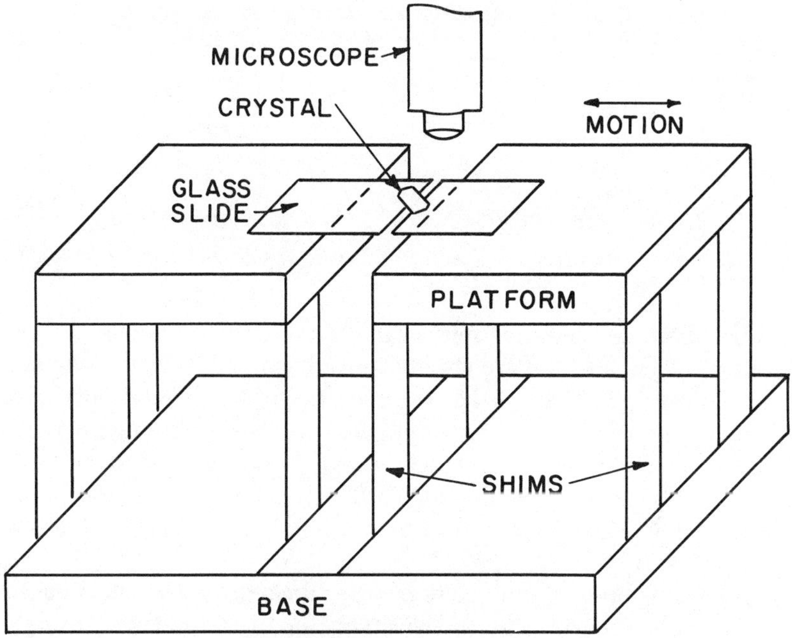

Figure 1. Diagram of nanotensilometer showing inverted pendula and crystal mounted on cracked glass slide

pendula, each supported by four flat beryllium–copper shims. This construction, shown in Figure 1, makes the following features possible: isolation from seismic disturbances; one-dimensional, effectively frictionless, motion; gage lengths from 1 μm to 1 mm; readily accessible sample mounting area; optical microscopy monitor during elongation; direct plotting of specimen force–elongation data; and hard and soft mode operation.

In this design the effective restoring force of the shims would ideally be less than that of the specimen. The instrument restoring force may be made arbitrarily small by adding mass to the platforms such that the platform resonant frequency approaches zero with the result that the force constant also approaches zero. In practice the platforms are not quite critically loaded and the small residual force is electronically subtracted so that the net force applied to a sample is directly plotted. Using gravitational loading to reduce the spring constant results in the adverse condition that the instrument is extremely sensitive to tilting of its base.

For operation in the hard mode, one selects the appropriate full-scale values of force, elongation, and the time to strain the sample. This predetermined strain rate signal is compared with an ac differential three-

Table I. Nanotensilometer Specifications[a]

Gage length	1–10^3 μm
Position range	1–10^6 nm
Force range	10^{-8}–5×10^{-1} N
Ramp time	10–10^6 sec
Strain rate	1–10^5 nm sec^{-1}
Hold time	> 10^6 sec
Sample cross-section	> 10^{-14} m^2

[a] Operates in both hard (position) and soft (force) modes; complete internal cancellation of machine force constant.

terminal capacitor position sensor and a dc current generated by the servocontrol circuitry. This control current energizes coils attached to each platform put in the field of Alnico magnets mounted on the base such that the relative force is applied to the specimen. Copper plates mounted on the sides of these coils provide eddy current damping. In the hard mode the nanotensilometer then separates the platforms at a predetermined strain rate and automatically adjusts the force to carry out this process.

Soft mode operation requires additional circuitry to apply a constant force to the specimen. The specifications for the tensilometer are given in Table I. For the position sensor used the precision is 0.1 nm, and the accuracy is 0.1% over the range 1–10^3 μm. The equivalent noise generated by the servocontrol is less than 10^{-8} N. The accuracy of the force output depends on the scale and the previous time-dependent deformation of the platforms and is approximately 2%. In practice values of the cross-sectional area and length of the specimen are the limiting features in transforming the data to stress–strain curves.

Sample Preparation and Mounting

The crystals have been solution grown in *o*-dichlorobenzene by Hasegawa and Geil. This technique and the resultant crystal morphology have been described by Geil (7). To prepare samples for tensile measurements several additional requirements must be met. First, the crystals must be dispersed in the solvent so no aggregates form, and the crystallization should be complete so that no amorphous debris remains. Furthermore, high purity solvents are necessary to avoid contamination of the gap upon evaporation of the solvent. In practice we have found it necessary to distill the solvents in several cases.

The sample holder is a section of glass microscope slide cracked in tension and beveled on its underside. An edge prepared in this way may be repeatedly closed to approximately 1 μm. Figure 2 is a photograph of a polyethylene (PE) crystal of planar dimensions 13×17 μm and 10 nm thick. As indicated in the photograph and schematically below, this crystal has been mounted across the gage length of 2 μm and then pulled to fracture. The photograph clearly indicates the crystal mounted on the microscope slides; as the photograph was taken after crystal fracture, no material is visible in the dark gap between the microscope slides. In practice visual observation with the microscope is difficult because of the small reflectivity of the crystals. PE is very light blue in reflected light when 10 nm thick.

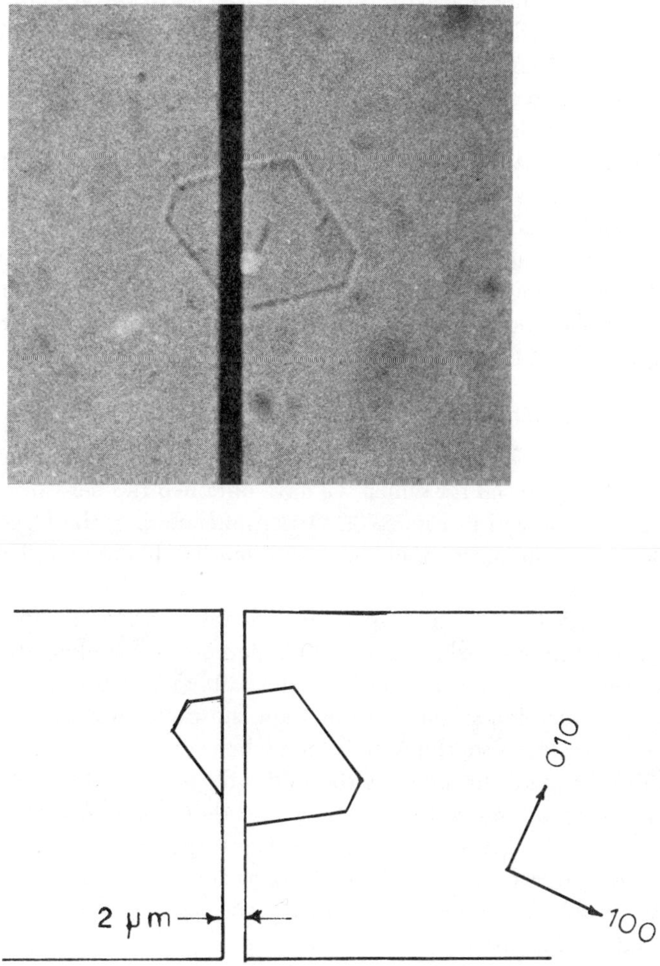

Figure 2. Photograph after fracture of PE crystal of planar dimensions 13 x 17 μm with gap and orientation as given in the lower diagram

The procedure for mounting the crystal consists of placing a small drop of the suspension containing the crystals on the gap between the two halves of the specimen holder. If a carefully selected solvent of an appropriate surface tension and vapor pressure is used, the crystals will not be damaged as the solvent evaporates. We have found isopropanol difficult, benzene usable, and *o*-dichlorobenzene best for PE. The two halves of the sample holder must be clamped rigidly to avoid motion of the jaws and damage to the sample during mounting. Crystals are approximately centered across the gap by the surface tension. The solvent serves the multiple function of transport and centering. If sufficiently clean, the crystal bonds electrostatically to the glass surface after the solvent evaporates. The positional jitter that results when the clamps are released is less than 2.5 nm.

In practice there are several difficulties. Small amounts of contamination by oils degrade the adhesion, and the crystals will slide rather than being stretched to failure. Because the solvent fills the gap between the two sides of the specimen holder, polymeric debris may be drawn into the gap and remain. Such bridges formed in this way are extremely difficult to observe but of course will contribute to the tensile data. At present the acquisition of tensile data is limited by the aggregation of crystallites and the noncrystalline material that bridges the gap. Crystals $< 10\ \mu m$ are difficult to handle.

Tensile Properties of Polyethylene Crystals

We illustrate the performance of the nanotensilometer with data for polyethylene, the material for which we have obtained the best individual crystal data as displayed in Figure 3a. Upon unclamping, the force rises in a pseudolinear fashion to a maximum of about 0.5 dyn at an elongation of about 100 nm when deformed at a strain rate of about 10 nm/sec with an initial gage length of 2 μm. Upon further deformation the force decreases in a continuous fashion but still is significant for elongations of several micrometers as can be seen in Figure 3b. Viscoelastic contributions are found for elongations greater than those corresponding to the peak in the force. Even so, the initial slope when reversing the strain rate remains approximately the same as the initial tensile curve.

Optical observations of the crystal while deforming indicates fractures occurring at elongations approximately corresponding to those of the peak of the force curve. During the decreasing force portion of the curve no optically visible specimen is seen across the gap. The most difficult part in operating this instrument is making certain that the material in the gap between the cracked microscope slide consists only of the desired single crystal specimen. To date we have obtained data for about 50 PE crystals or crystal aggregates.

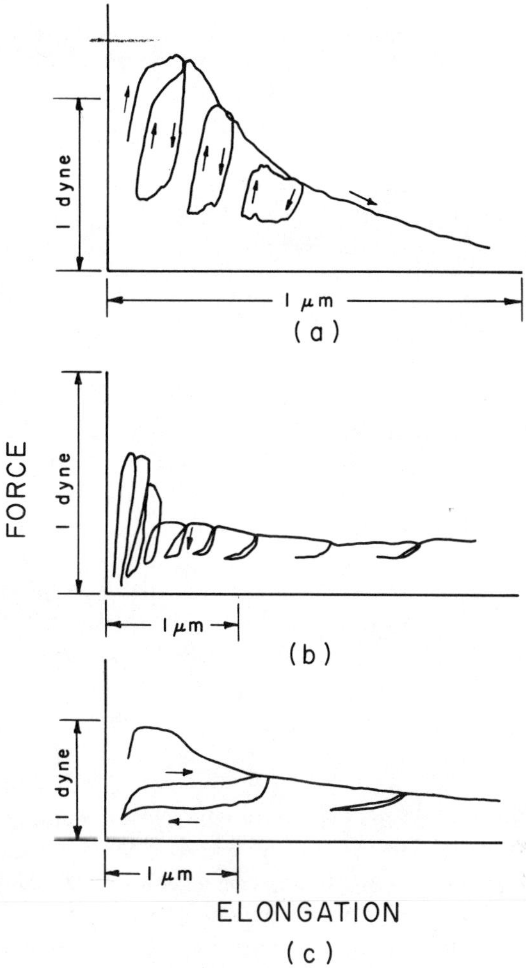

Figure 3. Force–elongation curves for PE crystals for gage length of about 2 μm. (a) Crystal of Figure 2; (b) crystal of different orientation; (c) average of crystals of arbitrary orientation.

We present these data on PE crystals as being qualitatively representative of the tensile properties of such crystals but recognize that the quantitative information must be regarded as preliminary. Not only must more accurate knowledge of the specimen cross-sectional area be obtained, but we must make certain the crystal surface is not contaminated with residue sufficient to affect the mechanical properties.

In spite of these difficulties we feel the following observations on PE deformed at room temperature will stand the test of time:

(1) The initial portion of the virgin force–elongation curve is approximately linear, and the crystals support compressional forces.

(2) Yield points are normally observed.

(3) Mechanical hysteresis observed at small strains shows little recovery.

(4) For strains larger than those corresponding to visual fracture only tensile forces are found and show high recovery from the extremely large (the order of 500%) extensions.

(5) Stress relaxation at large constant strain is observed.

These observations are illustrated in Figure 3c, showing the average tensile properties for four crystals of arbitrary orientation across an original 2-μm gage length.

In spite of our reluctance to quote numerical values at this point, the effective modulus obtained from the initial portion of the tensile curve ranges from 1 to 5×10^{10} dyn cm^{-2}. Many individual PE crystals have moduli from 3×10^{8} to 10^{10} dyn cm^{-2} and fracture at forces of about 0.2 dyn. Orientation effects are expected to be present and are presently being investigated. There is no comparable experimental data with which we can directly relate these values. However, moduli are in the range found for bulk specimens but are considerably less than the value of 70×10^{10} dyn cm^{-2} reported by Perkins et al. (*8*) for ultra-drawn HDPE fibers. The x-ray measurements of the lattice moduli by Ito (*9*) using an x-ray technique for oriented sheet samples is perhaps the most relevant comparison. He found values of 240×10^{10} dyn cm^{-2} in a direction parallel to the fiber axis and a value of 4×10^{10} dyn cm^{-2} for the perpendicular direction which would be the closest comparison with our orientation. We are not yet certain whether the initial portion of the stress–strain curve shows nonlinear viscoelastic effects such as found by Chen et al. (*4*) for springy polypropylene (PP) fibers.

We have had limited success with poly-4-methyl pentene-1 because the samples seem to consist of small crystals dispersed in a residue which conceivably could be amorphous polymer. However, we have now measured individual crystals of planar dimensions $< 10\ \mu$m. By comparison with the PE the poly 4-methyl pentene-1 samples were significantly more rigid and did not display the large recovery at high strains. In addition, the force–elongation curve was ragged, suggesting that a microdeformation process might be operating. A PE sample of the same appearance showed a similar tensile curve.

Discussion

The nanotensilometer is just beginning to produce useful information about single-crystal polymer specimens. In view of the difficulty of sample preparation and characterization at the present time, we must be quite speculative about the interpretation. Nevertheless, we suggest the following correlation with the physical processes as observed in the electron microscope (7). We feel the initial portion of the force–elongation curve corresponds to a stretching of the crystal with a modulus related to the interchain van der Waals forces and may correspond to an elastic stretching of the crystal for those strains. The existence of a force for an apparently fractured specimen is explained in terms of stretching of microfibrils across the gap as the specimen is deformed. The diameter of such fibrils is of the order of 10 nm[7] and hence would not be observed under the optical microscope. At present we are not able to count the number of such fibrils to determine an individual modulus. However, we suggest that since a substantial portion of the force remains supported to sizeable elongations, individually each fibril may possess a high strength. In both the crystal modulus measurements as well as in the details of the stretching, orientation effects are expected to be important and are observed by electron optics. We are optimistic that this new technique will provide a connection between the electron microscope deformation studies and the macroscopic tensile data that have been available. Indeed, we suggest that the change in character from the small-strain to large-strain behavior observed in PE may be related to the drawing process to produce high modulus fibers.

Acknowledgments

This research was part of the interdisciplinary activity at CWRU supported by the NSF Materials Research Laboratory Program. We gratefully acknowledge the contributions of P. H. Geil and H. Hasegawa in growing the polymer crystals, in determining the polymer morphology, and for many valuable discussions. We also thank Carl Hagerling for the specimen mounting and for operation of the nanotensilometer.

Literature Cited

1. Hoffman, R. W., "Physics of Thin Films," Thun and Hass, Eds., Vol. 3, p. 211, Academic, New York, 1966.
2. Brenner, S. S., *J. Appl. Phys.* (1956) **27**, 1484.
3. Riley, M. W., Skove, M. J., *Phys. Rev. B* (1973) **8**, 466.
4. Chen, H. Y-L., Ko, F. K., Lundberg, J. L., *Polym. Eng. Sci.* (1976) **16**, 406.
5. Fields, R. W., *Biophys. J.* (1970) **10**, 462.
6. Andeen, C. G., Hagerling, C. W., Hoffman, R. W., *Proc. Int. Vac. Congr., 7th,* and *Int. Conf. Solid Surfaces, 3rd* (Vienna 1977) (1977) **2**, 1769.
7. Geil, P. H., "Polymer Single Crystals," Wiley Interscience, New York, 1963.
8. Perkins, W. G., Capioti, N. J., Porter, R. S., *Polym. Eng. Sci.* (1976) **16**, 200.
9. Ito, T., Memoirs of Faculty of Ind. Arts, Kyoto Tech. Univ., *Sci. Technol.* (1966) **15**, 43.

RECEIVED November 10, 1977.

4

Measurement of Nonlinear Viscoelastic Properties of Polymers in Cyclic Deformation under a Relatively Large Strain Amplitude

Y. D. KWON, R. K. SHARMA, and D. C. PREVORSEK

Chemical Research Center, Allied Chemical Corp., Morristown, NJ 07960

An apparatus and method for measuring nonlinear viscoelastic properties of polymers in cyclic deformations under relatively high strain amplitude ($\pm$ 0.1 ~ 2%) are described. By superimposing a high frequency, small amplitude sinusoidal strain on a low frequency, high amplitude fundamental sinusoidal strain, the apparatus allows the determination of instantaneous modulus as a function of the phase angle during a cycle. Results of measurements are analyzed in terms of the phase angle difference between the nonlinear elastic stress and the nonlinear viscoelastic stress. Three major factors—the nonlinear elasticity, strain-rate dependence of the frictional viscosity, and the reversible structural change induced by the cyclic straining—are believed to contribute to the observed stress response. Modes of their contributions are discussed.

In many practical applications of viscoelastic materials under cyclic straining conditions the strain amplitude which the material experiences in the applications is too large to allow the assumption of linear viscoelasticity. For example, the tire cord experiences a strain amplitude of about 1% or more (1) while the tire for a passenger car runs on the road under a normal condition. Under the strain amplitude of this magnitude viscoelastic behavior of the material deviates from linearity significantly, and therefore analysis of the viscoelasticity must consider the nonlinearity.

0-8412-0406-3/79/33-174-035$05.00/1

In a previous article (2) we described a new approach to the analysis of nonlinear viscoelasticity as encountered in such large-strain cyclic deformation of polymers. In this chapter, we describe the apparatus used and the method of treating the data obtained.

Measurements of viscoelastic properties under severe deformation which may lead ultimately to rupture have considerable technological and theoretical importance. For example, in periodic deformation the lifetimes of a set of specimens from a sample show large scatter. With fibers it is not unusual that a section of a fiber which was broken in a fatigue experiment has a lifetime which is comparable with or higher than the lifetime of the original specimen.

These and similar results led to the formulation of the statistical theories of fiber endurance and strength which assume that the ultimate properties of fibers are controlled by the properties of the weakest cross section of the specimen. The problem with this theory is that it does not lead to an estimate of the ultimate properties of a flawless specimen. Consequently, it is impossible to speculate at present about the strength and endurance of specimens manufactured under ideal conditions from the properties of a given sample.

Another important and unsolved problem of polymer mechanics is determination of the residual life of a sample which was subjected to periodic straining short of rupture. Previous studies showed that the breakdown of oriented polymers is catastrophic and that failure is not preceded by a gradual decay of specimen strength.

It appears, therefore, that measurements of strength do not reveal the important characteristics of structural cleavages which occur during fatiguing and which ultimately lead to rupture. Only a few isolated experiments were done in which the viscoelastic properties were determined during fatigue experiments. Nevertheless, the results look promising. For example, Nielsen reports (3) that glass-reinforced nylon specimens show a marked increase in tan δ and a decrease in modulus shortly before rupture. Since a semicrystalline polymer can be considered a two-phase composite, we speculated that similar changes in mechanical losses and modulus could be observed also with nonreinforced semicrystalline fibers. We decided therefore, to undertake a systematic study of dynamic viscoelastic properties under conditions leading to rupture.

In the course of this work we realized that the available instruments are not suitable to carry out simultaneously mechanical and rheological measurements. For example, instruments which are suitable for determining polymer transitions operate at low-strain amplitudes and thus cannot be used to carry out fatigue experiments to rupture polymer specimens of normal and medium strength. In addition, most of the rheological instruments cannot measure the viscoelastic properties if the stress resulting from a sinusoidal strain is not sinusoidal (i.e., with nonlinear viscoelastic materials).

Furthermore, we found that under severe cyclic deformation, the specimen dimensions, structure, and properties change considerably as a function of time. In order to conduct meaningful viscoelastic experiments it was necessary therefore, to carry out continuously the measurements of creep and control pretension automatically.

Another problem which arises in a cyclic deformation under a large strain amplitude is that the modulus of material varies significantly during a cycle because of strain-stiffening or strain-loosening effects. Therefore, analysis of the viscoelastic properties in this case must include the variation of modulus during a cycle.

Recent studies show that nonlinear stress response can be represented in terms of a time-dependent phase angle difference δ between the nonlinear elastic stress and viscoelastic stress. This approach appears to have an advantage over the numerical treatments used by previous workers because it provides some information about the reversible structural changes which occur during the cycle. However, such analyses require instruments which can subject the specimen simultaneously to two cyclic deformations of different amplitude and frequency.

The more versatile instruments are equipped to carry out simultaneously the fatigue and nonlinear viscoelastic experiments. Nevertheless, the use of these expensive instruments is not recommended for long-term fatigue experiments in which a specimen may have to be cycled for several days or weeks under same conditions. We decided, therefore, to construct an instrument which combines the essential characteristic of the low strain amplitude Rheovibron with those of sophisticated fatigue apparatus. The apparatus described in this chapter has the following capabilities compared with the commercially available rheological and fatigue instruments:

(1) application of a sinusoidal strain to a polymer specimen with an amplitude of 0.1 $\sim$ 2.0% under tension, compression, or shear in the frequency range between 0.1–50 cycles/sec between −30°C and 250°C;

(2) determination of the modulus $E(\theta)$ as the function of phase angle θ by superimposing a smaller amplitude, high frequency strain on the fundamental strain. From this result, the phase angle difference between the elastic stress and viscoelastic stress can be determined;

(3) display of hysteresis loop, time variaitons of stress and strain on a common time base, and modulus of the specimen;

(4) automatic integration and display of the hysteresis loop area;

(5) determination of creep as a function of time; and

(6) determination of fatigue life of a specimen under cyclic deformation.

Apparatus

A schematic of the apparatus is shown in Figure 1. It consists of two variable speed motors, one connected to an eccentric for generating a small sinusoidal strain and the other for generating the fundamental sinusoidal strain. The eccentric has a slot in which an insert can be moved diametrically across the face of the eccentric and locked into a position, permitting the adjustment of the stroke to a selected strain amplitude. The insert has posts 1, 2, and 3 (Figure 1). Posts 1 and 2 and the center of driving shaft are colinear, and the line connecting the post 3 and the center of the shaft is perpendicular to the line formed by the first two. Each of the posts 2 and 3 is connected via a rigid linkage to a spring which is further connected to a transducer (strain transducer) and a readout unit. Post 1 is attached by a rigid linkage to the left-hand side of the sample. The right-hand side of the sample, which is shaped into a loop when tension is applied to the fibers, is attached to another rigid linkage, to a transducer (stress transducer), and to the stress readout unit. The two strain transducers are mounted on micropositioners which can be adjusted finely to achieve a colinearity between the right-hand (stress) and the left-hand (strain) transducers, and the center of the driving shaft, and to form 90° angle between the line formed by the vertical strain transducer and center of the driving shaft and the line formed by the first two transducers. The stress transducer is mounted on a movable carriage; its motion being controlled by a servomotor so that the

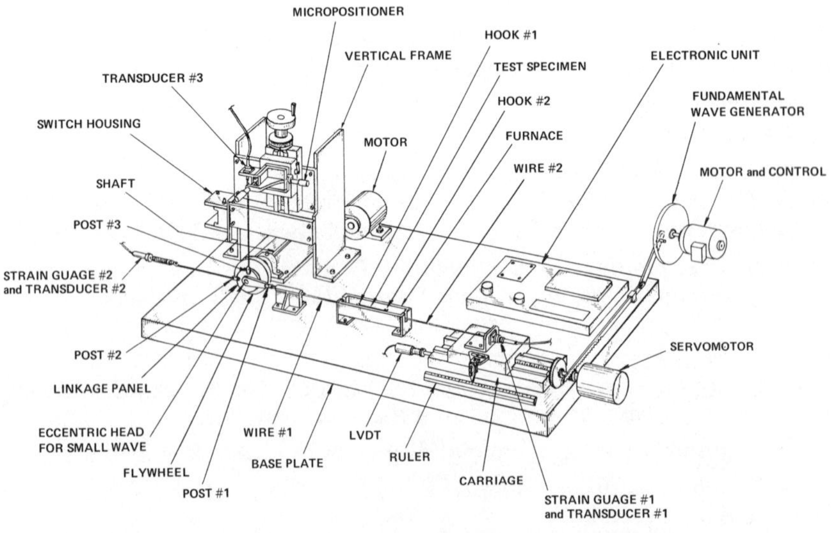

Figure 1. Schematic of the apparatus

sample is always maintained at a constant predetermined median pretension. When the fundamental strain generator is driven, it generates a sinusoidal, reciprocating motion of the carriage on which the strain gage 1 is positioned, thus generating the sinusoidal fundamental strain. The movement of this carriage is monitored by a linear variable differential transformed (LVDT) so that any extension or shrinkage of the sample during the experiment can be measured accurately. The sample is enclosed in a furnace whose temperature can be controlled to a desired level with an accuracy of $\pm 0.5°C$ in the range of $-30°$ to $250°C$.

The signals from the stress and left-hand strain readout units can be displayed on an oscilloscope simultaneously against time to monitor the sinusoidal strain signal and the stress signal which is approximately sinusoidal. Or the stress signal can be displayed against the strain signal to form the hysteresis loop. The loop can be photographed using a camera and the area within the loop, representing the loss per cycle, measured. Alternatively, the stress signal and strain signal from the vertical strain transducer, which represents the first derivative of the strain signal from the left-hand transducer, can be fed into the electronic integrator. The integral signal can be displayed either on the oscilloscope and photographed or on the digital voltmeter. When the small strain is superimposed onto the fundamental strain, the resulting composite stress can be displayed on the oscilloscope or photographed from the oscilloscope display.

Operation of the Apparatus

Operation of the apparatus for an actual measurement requires the following steps.

Calibration of the Strain Gages, Transducers, and Signal Conditioners. The stress transducer and signal conditioner system is calibrated by determining the coefficient C_σ [volts/(kg/cm^2)] which represents the force-to-volts ratio of the system. This is accomplished by applying a known static force (such as a weight) and recording the output of the system. Similarly, both the strain transducer and signal conditioner systems are calibrated by determining the coefficients C_γ and $C_{\dot{\gamma}}$ [volts/(cm/cm)] which represent the volt-to-strain ratio of these two systems. This is accomplished by imposing a known displacement to the transducers across the attached springs and recording the outputs of the systems. The transducer (for example, Statham's Universal Transducing Cell model UC2) is connected to a load cell (for example, Statham's Load Cell Accessory model UL4-100). The load cell responds to the applied force by a small displacement of its core rod, the displacement being proportional to the force, and is transferred to the transducer. The

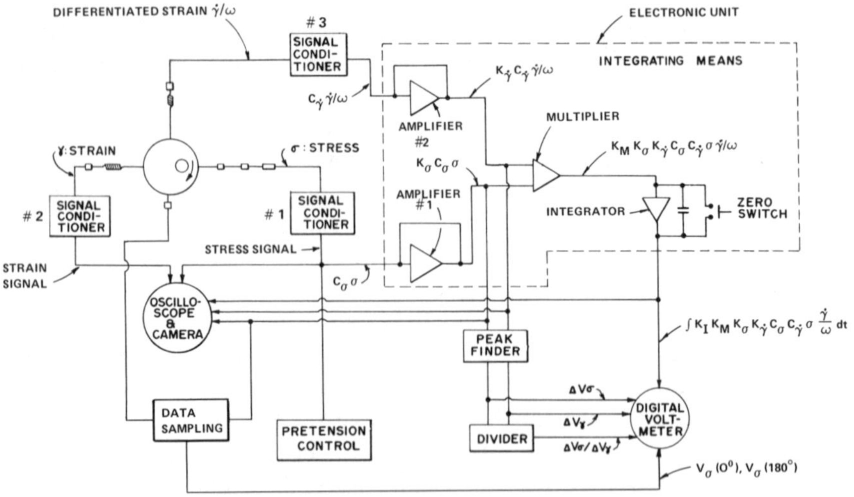

Figure 2. Electronic system

transducer then converts this displacement to a proportional voltage output. Next, amplification factors of the amplifiers 1 and 2 (Figure 2) are set to desired values. These factors, K_σ and $K_{\dot{\gamma}}$, respectively, represent how many volts are put out by the amplifier when one volt is fed into it. Similarly, the factors, K_M and K_I for the multiplier and integrator, respectively, are also adjusted.

Alignment of the Relative Positions of the Transducers. The stress transducer (transducer 1), the strain transducer (transducer 2), and the center of the drive shaft (Figure 1) must be positioned colinearly. This is done first by the use of the micropositioner to which the strain transducer is attached. Then for an accurate alignment, the hysteresis loops of Figure 3 are generated with the drive shaft rotated in the forward and

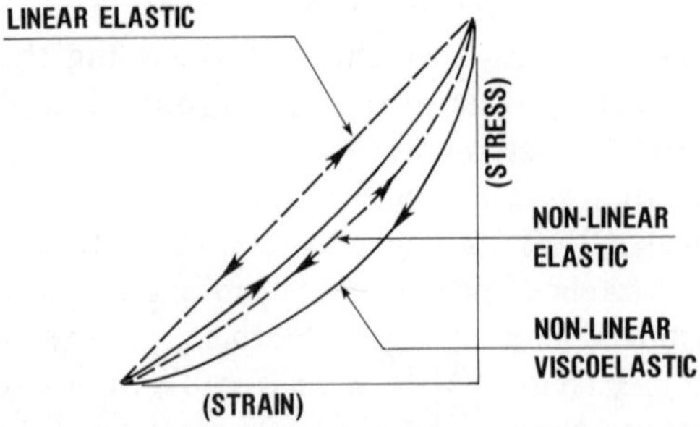

Figure 3. Stress–strain curves in nonlinear elastic and nonlinear viscoelastic responses

reverse directions, finely adjusting the altitude of transducer 2 such that the two resulting loops are identical. Transducer 3 is positioned (Figure 1) such that the line connecting it to the center of the drive shaft is perpendicular to the line connecting transducers 1 and 2. This can be done by one of the two following methods.

(1) Install a spring in place of the test specimen and adjust the amplitudes of the strain output from transducers 2 and 3 to an equal value. Display them on the oscilloscope one against the other. When the alignment is accurate, the display should generate a circle; any observable deviation of the generated circle from a perfect circle indicates a misalignment.

(2) Display the two strain outputs simultaneously on the time base and ascertain that the phase difference between the two sinusoidal curves is 90°.

Control of the Air Temperature Surrounding the Specimen. The air temperature in the constant temperature furnace (Figure 1) is controlled to a desired level in the range of $-30°$ to 250°C by the temperature controller with an accuracy of ± 1°C. Differences between the specimen temperature and the air temperature depends on the heat generation rate in the specimen during the cyclic straining and the convective heat transfer coefficient for the heat transfer between the specimen and the air surrounding it.

Thus, at a steady state equilibrium between the specimen temperature T_s and the air temperature T_a the following relationship holds:

$$\pi DL \cdot h_c \cdot (T_s - T_a) = \dot{q} \frac{\pi D^2}{4} L$$

where D is the filament diameter, L is the filament length, q is the heat generation rate per unit volume of the specimen, and h_c is the convective heat transfer coefficient. The difference between the sample temperature and the chamber temperature can amount to several degrees, depending upon the experimental conditions. This is always taken into account using the equation shown above.

Setting of Pretension, Strain Amplitude, and Frequency of Cyclic Straining. Pretension can be controlled by using the pretension controller (e.g., Honeywell Servo Amp. 369960-103) connected to the servomotor (e.g., 11.5 W 105 oz-in., Honeywell Servomotor). When desired, the pretension control can be carried out manually. Strain amplitude is set by using the adjusting screw to move the slide in the desired direction (Figure 4). The frequency of cyclic straining is set by using the motor-speed controller attached to the drive motor (Figure 1). The frequency can be determined accurately by putting the strain signal on the time base of the oscilloscope and measuring the peak-to-peak distance.

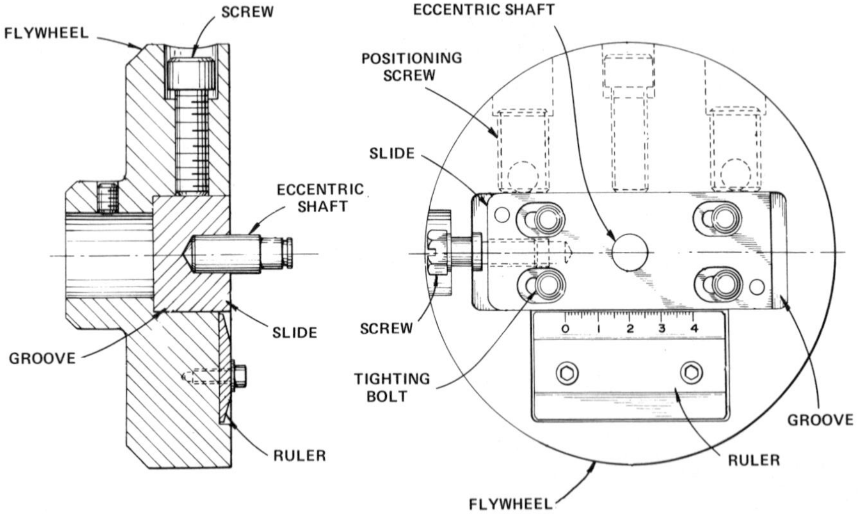

Figure 4. Details of the flywheel

The frequency and strain amplitude of the fundamental strain, in the case of superimposing the small strain for modulus measurement, are adjusted by use of the motor control and the cam setting attached to the driving system.

Conditioning of the Specimen. When subjected to large strain amplitudes, the specimen initially undergoes a considerable degree of structural change and gradually reaches an equilibrium. This is manifested in the variation of specimen length (creep), the mechanical loss, and modulus of the specimen. These variations will be discussed in a subsequent paper. When the data for the steady state are to be obtained, the specimen has to be conditioned for a long time.

Outline of Data Analysis

The principles of analysis as applied to actual experimental data and the definitions of various terms and concepts involved are discussed in detail elsewhere (2).

Basic Approach. We consider the cases in which a viscoelastic specimen such as a piece of nylon monofilament is subjected to a sinusoidal straining expressed by:

$$\gamma(t) = \gamma_o + \Delta\gamma \sin \omega t \tag{1}$$

where γ_o is the median strain, $\Delta\gamma$ is the strain amplitude, ω is the frequency, and t is time. $\Delta\gamma$ is on the order of $\pm 1\%$.

When $\Delta\gamma$ is very small ($\Delta\gamma \leq 0.1\%$), the stress response caused by the sinusoidal straining given by Equation 1 is approximately sinusoidal, and the viscoelastic behavior falls in the region of linear viscoelasticity. In this case, the phase angle difference δ between the stress wave and strain wave is constant throughout the cycle, and the stress response can be expressed by:

$$\sigma(t) = \sigma_0 + \Delta\sigma \sin(\omega t + \delta) \tag{2}$$

There have been extensive studies of viscoelastic behavior in this linear region (*see* Ref. *4*, for example).

When $\Delta\gamma$ is relatively large as in the present case, $\sigma(t)$ is not sinusoidal, and the viscoelastic behavior falls in the nonlinear region. In this case, previous workers analyzed the stress response by a Fourier series representation (*5*). Results of such analyses are useful for storing and reproducing the data but are not useful in the physical interpretation of the changes occurring in the material within a cycle during the cyclic straining.

When the strain amplitude in a sinusoidal straining is relatively large, the elastic modulus of the specimen varies significantly within a cycle because of the strain-stiffening or strain-softening phenomena. Therefore, the first step of the analysis is to determine this variation of modulus as a function of the strain. From the modulus data, the nonlinear elastic stress response corresponding to the sinusoidal straining of Equation 1 can be established. Then, by comparing the nonlinear elastic stress response with the actual viscoelastic stress response, one can determine the phase angle difference, δ, which reflects the net effect of the nonelastic contribution. The value of δ varies as a function of the phase angle ωt (or θ). Thus, the nonlinear viscoelastic stress response can be decomposed into the elastic component and nonelastic component and the results interpreted in terms of the structural changes occurring during the sinusoidal straining.

Characteristics of the Hysteresis Loop and Stress Wave in the Nonlinear Viscoelastic Response to the Sinusoidal Straining. Figure 3 is a schematic of a hysteresis loop obtained when a nylon 6 monofilament was subjected to a sinusoidal straining with $\gamma_0 = 1\%$ and $\Delta\gamma = \pm 1\%$ at 90°C under a frequency of 10 cycles per sec.

If this specimen were behaving in a linear viscoelastic pattern, the loop would be an ellipse whose long axis lies on the linear elastic line indicated by the dotted line. The experimental hysteresis loop in Figure 3, however, is bent downward considerably. The extent of this downward bending is such that even the upper curve of the loop which represents the stretching phase is sometimes located below the linear elastic line.

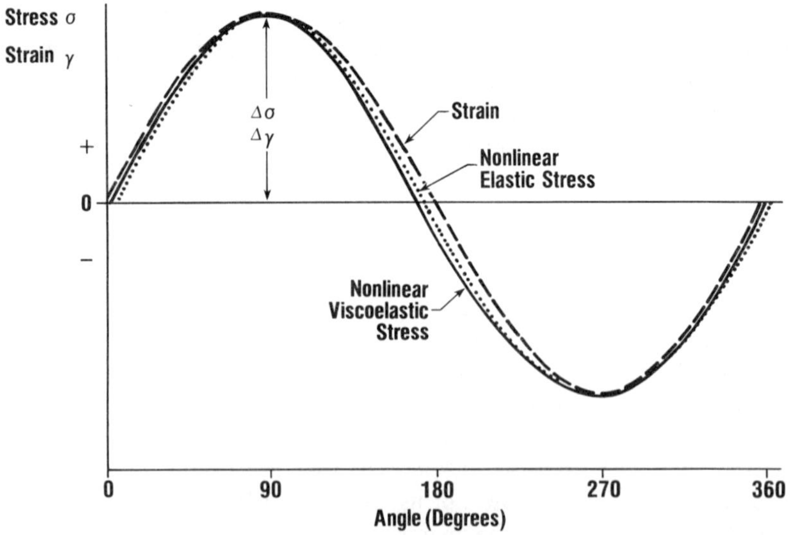

Figure 5. Variation of stress and strain with phase angle during a cycle

One reason that the loop curves are bent downward so that the slope increases as the strain increases is obviously the strain-stiffening effect. As the specimen is stretched beyond the linear elastic limit, the modulus of the material and hence the slope of the curve increases with increasing strain.

Figure 5 is a schematic of the sinusoidal strain and the corresponding viscoelastic stress response displayed together against a common time base (as seen on a dual display oscilloscope). The figure also shows the nonlinear elastic stress which is calculated from the modulus. The most interesting observation in this display is that the phase angle difference between the stress curve and strain curve (which is sinusoidal) varies in magnitude as well as in sign as functions of the phase angle. In this display the amplitudes of stress and strain were made equal by proper scaling and shifting of position.

A part of this peculiar variation in the stress–strain phase angle difference comes from the variation of elastic modulus with the phase angle. Therefore, to determine the phase angle difference which is caused by the nonelastic contribution, it is necessary to determine and to subtract the contribution from the variation of elastic modulus as a function of the strain.

Determination of Elastic Modulus Varying as a Function of Strain during a Cycle. To determine the elastic modulus which varies as a function of strain, we superimpose a small amplitude, high frequency

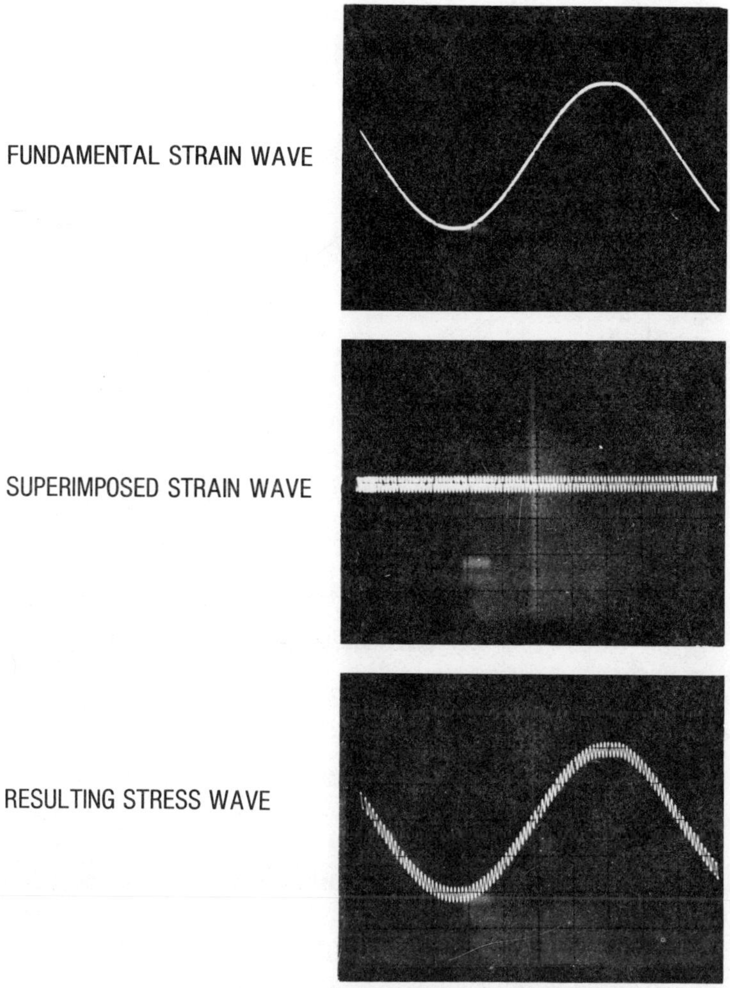

Figure 6. Stress response to the fundamental strain wave superimposed with smaller amplitude strain wave. (top) *Fundamental strain wave;* (middle) *superimposed strain wave;* (bottom) *resulting stress wave.*

strain (superimposed strain) onto the large amplitude, low frequency strain (fundamental strain).

Figure 6 is an example of the resulting stress response. The modulus at a phase angle θ can be obtained by dividing the superimposed stress stroke $(\Delta\sigma)_s$ by the corresponding superimposed strain stroke $(\Delta\gamma)_s$:

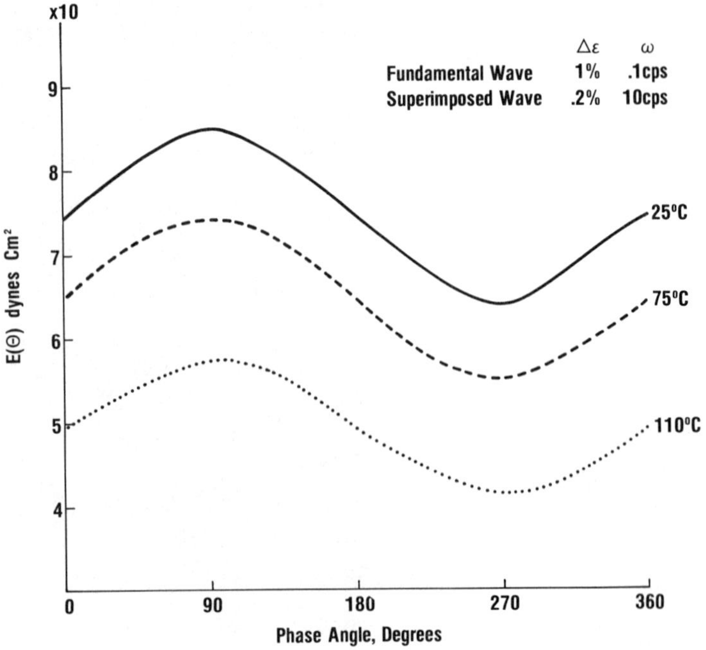

Figure 7. E(θ) of nylon 6 monofilament

$$E(\theta) = \left.\frac{(\Delta\sigma)_s}{(\Delta\gamma)_s}\right|_\theta \quad (3)$$

Figure 7 is an example of how this elastic modulus varies as a function of the phase angle θ and of the strain. In this case, the modulus varies almost sinusoidally with respect to the phase angle, indicating the strain-stiffening behavior of the sample.

In view of Equation 1, the elastic modulus in this case can be expressed by a sinusoidal form of:

$$E(\theta) = E_o + \Delta E \sin \theta \quad (4)$$

or as a linear function of strain:

$$E(\gamma) = E_o + \Delta E'(\gamma - \gamma_o) \quad (4a)$$

Nonlinear Elastic Stress Response to the Sinusoidal Straining. When a nonlinear elastic body whose modulus varies by Equation 4 is subjected to cyclic straining, the stress response would be:

$$\sigma_{el}(\gamma) = \sigma_{el}(\gamma_o) + \int_{\gamma_o}^{\gamma} E(\gamma)\,d\gamma \tag{5}$$

Referring to Figure 5, we can take the minimum strain point as the reference point and apply Equation 5 to establish the nonlinear elastic stress response $\sigma_{el}(\gamma)$ to the sinusoidal straining:

$$\sigma_{el}(\theta) = \sigma_{\min} + \int_{3\pi/2}^{\theta} E(\theta) \cdot \Delta\gamma \cdot \cos\theta\, d\theta \tag{6}$$

When $E(\theta)$ is sinusoidal as given by Equation 4, $\sigma_{el}(\theta)$ can be integrated analytically. When $E(\theta)$ is not sinusoidal, integration can be done numerically.

In the case where $E(\theta)$ is given by Equation 4, Equation 6 yields:

$$\sigma_{el}(\theta) = \sigma_{\min} + E_o\,\Delta\gamma - \frac{\Delta E}{2}\,\Delta\gamma + \Delta\gamma\left[E_o \sin\theta + \frac{\Delta E}{2}\sin^2\theta\right] \tag{7}$$

or

$$\sigma_{el}(\theta) - \sigma_{\min} = E_o\,\Delta\gamma\left[1 + \sin\theta + \frac{e}{2}(\sin^2\theta - 1)\right] \tag{8}$$

where

$$e = \Delta E / E_o.$$

The maximum value of elastic stress should correspond to $\theta = 90°$ where the strain is at its maximum. Therefore:

$$\sigma_{el_{\max}} - \sigma_{\min} = 2E_o\Delta\gamma \tag{9}$$

When we use the value of E_o obtained with the relatively small amplitude of the superimposed strain $(\Delta\gamma)_s$, $\sigma_{el_{\max}} - \sigma_{\min}$ from Equation 9 is larger than the experimental value of $\sigma_{d_{\max}} - \sigma_{\min}$ (where σ_d indicates the viscoelastic stress). This is caused by the fact that the modulus observed with a high frequency, smaller strain amplitude is larger than that observed with a low frequency, larger strain amplitude. Assuming that the actual $\sigma_{el_{\max}}$ is approximately equal to $\sigma_{d_{\max}}$, we adjust the values of $E(\theta)$ by a factor f given by Equation 10:

$$f = \frac{(\sigma_{d_{\max}} - \sigma_{\min})_{\text{exp.}}}{(\sigma_{el_{\max}} - \sigma_{\min})_{\text{calc.}}} \tag{10}$$

Equation 8 is modified to:

$$\sigma_{el}(\theta) - \sigma_{\min} = f\,E_o\Delta\gamma\left[1 + \sin\theta + \frac{e}{2}(\sin^2\theta - 1)\right] \tag{11}$$

The term $\sigma_{el}(\theta)$ given by Equation 11 appears as a single curve for the stretching and contracting phases in the stress–strain plane. This is shown by the dotted curve with two arrows in opposite directions in Figure 3. The elastic stress is also shown by the thinner dotted line in Figure 5. That this elastic stress response appears by a single line on the stress–strain plane is only natural because the elastic stress response should not involve any hysteresis loss.

In Figure 5, the elastic stress curve falls between the strain curve and the viscoelastic stress curves during the contracting phase, showing that a part of the phase angle difference between the strain and viscoelastic stress is contributed by the nonlinear elasticity of the specimen.

Phase Angle Difference between the Nonlinear Viscoelastic Stress and Nonlinear Elastic Stress. The phase angle difference δ between the nonlinear viscoelastic stress and nonlinear elastic stress is defined by the relation:

$$\sigma_d(\theta) = \sigma_{el}(\theta + \delta(\theta)) \tag{12}$$

This phase angle difference represents the net effect of the nonelastic contribution, and its physical significance corresponds to that of constant δ measured for the linear viscoelastic material.

Having obtained the values of $\sigma_{el}(\theta)$ and $\sigma_d(\theta)$, the values of $\delta(\theta)$ as defined by Equation 12 can be determined graphically or numerically. Graphical determination of $\delta(\theta)$ is difficult when the values are small. A more efficient numerical procedure is to start from the hysteresis loop of Figure 3 and to proceed as follows.

First, either by visual reading or by computerized data acquisition system, read off the coordinates of σ_d and γ corresponding to each other. Since γ is controlled to the sine form of Equation 1, it is easy to determine the phase angle difference $\delta'(\theta)$ between σ_d and γ as defined by:

$$\sigma_d(\theta) = \sigma_o + \Delta\sigma \sin(\theta + \delta'(\theta)) \tag{13}$$

Now, Equations 11, 12, and 13 can be combined to yield:

$$\sigma_{\min} + fE_o\,\Delta\gamma\left[1 + \sin(\theta + \delta(\theta)) + \frac{e}{2}(\sin^2(\theta + \delta(\theta)) - 1)\right] = \sigma_o + \Delta\sigma\sin(\theta + \delta'(\theta)) \tag{14}$$

Values of $\delta(\theta)$ can be obtained by solving Equation 14.

Variation of $\delta(\theta)$ as a Function of Phase Angle and Strain Amplitude and Its Physical Implications. Figure 8 shows an example of the variation of $\delta(\theta)$ as a function of the phase angle and strain amplitude determined for nylon 6 fibers at 90°C. In this figure, we make the following observations.

First, $\delta(\theta)$ varies periodically, and the peak values are in the vicinity of $\theta = 0°$ and 180°. This suggests that $\delta(\theta)$ is in phase with the strain rate. The position of the peak value shifts with the change of strain amplitude.

Second, as the strain amplitude increases, the peak value of $\delta(\theta)$ in the contracting phase becomes higher relative to the peak value in the stretching phase. This fact points to the interesting possibility of a reversible change in the polymer structure associated with reversible release and absorption of energies. If $\delta(\theta)$ only indicates the energy loss caused by internal friction in the cyclic straining, $\delta(\theta)$ during the stretching phase should be symmetrically equal to that during the contracting phase. The fact that this is not so implies that there is a cyclic energy interaction which is not caused by the internal viscous friction. And the degree of this energy interaction is larger when the strain amplitude is larger.

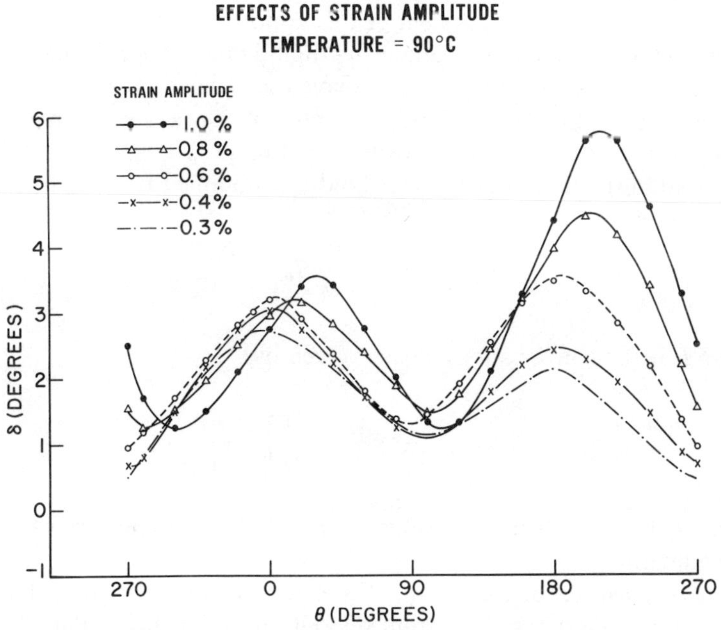

Figure 8. Variation in δ during cycle (nylon 6 fibers). Temperature = 90°C.

At this time, we do not know the exact nature of this energy interaction. However, we speculate that mechanisms which can account for such effects may be strain-induced crystallization, orientation, etc. upon stretching. Thus, during the stretching phase, strain-induced crystallization would cause a release of energy from the polymer, and during the contracting phase, this energy would be absorbed back.

Attempts will be made to ascertain this consideration by dynamic x-ray measurements of the crystallinity during the cycle. But the task may be difficult because of the small magnitude of the crystallinity change involved during the cyclic straining (on the order of 0.1%). Further details of the variation of $\delta(\theta)$ with temperature and other parameters are discussed elsewhere (*6*).

Automatic Integration of the Hysteresis Loop. The hysteresis loss of the specimen per cycle corresponds to the area of the loop shown in Figure 3. Thus, determining the hysteresis loss E_d (energy dissipated per cycle per unit volume) amounts to finding the integral:

$$E_d = \int_0^{2\pi} \sigma_d \, d\gamma \tag{15}$$

E_d can be obtained by the graphical integration of the loop area. Using the automatic integrating system of this apparatus, however, the integral can be obtained by simply reading the digital voltmeter (Figure 2). The voltage from the integrator as displayed on the digital voltmeter V_d corresponds to the accumulation of integral for a certain number of cycles n as set by the cam switch of Figure 1. E_d is conveniently expressed in the unit of erg/cc/sec, and it is obtained from V_d by:

$$E_d = \frac{V_d}{n} I_c \tag{16}$$

where I_c is the conversion coefficient given by:

$$I_c = \frac{1}{C_\sigma C_{\dot{\gamma}}} \cdot \frac{\omega}{v} \cdot \frac{9.8 \times 10^5}{K_\alpha K_{\dot{\gamma}} K_M K_I} \tag{17}$$

where v is the specimen volume in cc. Other parameters have been defined earlier.

Measurements of the Creep. The creep of samples during the cyclic straining can be determined from the output of LVDT (Figure 1). The output needs to be converted to length unit by a predetermined calibration factor.

Representation of the Nonlinear Dynamic Viscoelasticity in Terms of the Effects of Nonlinear Elasticity, Strain-Rate-Dependent Viscosity, and Reversible Strain-Induced Structural Change

The above analysis shows that the nonlinear dynamic viscoelastic behavior of polymers can be resolved into three components: the nonlinear elasticity resulting from the variation of modulus with the phase angle or strain during the cycle; nonlinear internal friction resulting from strain and strain-rate dependence; and effects associated with the reversible, strain-induced structural changes.

According to this representation, the nonlinear viscoelastic stress response resulting from sinusoidal strain:

$$\gamma(\theta) = \gamma_o + \Delta\gamma \sin \theta \tag{18}$$

can be represented by Equation 12. $\sigma_{el}(\theta)$ in Equation 12 is given by Equation 6. The phase angle difference $\delta(\theta)$ is separated into two parts: the part associated with the strain and strain-rate dependent, frictional viscosity δ_{vis} and the part associated with the effect of reversible strain-induced structural change $\Delta\delta$. Thus,

$$\delta(\theta) = \delta_{vis}(\theta) + \Delta\delta(\theta) \tag{19}$$

A quantitative analysis of stress waves in terms of nonlinear elastic, nonlinear viscous, and reversible strain-induced energy effects will be presented in a subsequent publication.

Conclusions

The apparatus described in this article can be used to measure the modulus and the elastic stress–viscoelastic stress phase angle difference as functions of the phase angle in a sinusoidal straining of a specimen and to obtain the hysteresis loss through an automated integral circuit.

The angular dependence of the elastic–viscoelastic stress phase angle difference $\delta(\theta)$ reveals a heretofore unnoticed pattern of variation during a cycle, i.e., a periodic change of its value $\delta(\theta)$ during stretching is smaller than in contracting with nylon 6 monofilament. This brings a very important implication to the interpretation of the data in terms of structural changes which occur during the sinusoidal straining.

Literature Cited

1. Prevorsek, D. C., Kwon, Y. D., Sharma, R. K., "Method of Determining Dynamic Strains in Composite Structures," U. S. Patent No. **3,934,452**, January 1976.

2. Prevorsek, D. C., Kwon, Y. D., Sharma, R. K., 'Interpretive Nonlinear Viscoelasticity: Dynamic Properties of Nylon 6 Fibers," *J. Macromol. Sci., Phys.* (1977) **B13**(4), 571.
3. Nielsen, L. E., *J. Compos. Mater.* (1975) **9**, 149.
4. Ferry, J. D., "Viscoelastic Properties of Polymers," Wiley, New York, 1970.
5. Matsumoto, T., et al., "Nonlinear Behavior of Viscoelastic Materials, II. The Method of Analysis and Temperature Dependence of Nonlinear Viscoelastic Functions," *Trans. Soc. Rheol.* (1973) **17**(1), 47.
6. Prevorsek, D. C., Sharma, R. K., Kwon, Y. D., "Strain Induced Structural Charges in 'Hard Elastic' Polypropylene," *Am. Chem. Soc., Div. Org. Coat. Plast. Chem., Prepr.* (1978) **19**(1), 332.

RECEIVED November 10, 1977.

5

Stress Mass Spectrometry of Polymeric Materials: A Review

M. A. GRAYSON and C. J. WOLF

McDonnell Douglas Research Laboratories, McDonnell Douglas Corp., St. Louis, MO 63166

The instrumental aspects and applications of stress mass spectrometry (stress MS) to polymeric materials is reviewed critically from the inception of the technique to the present. Stress MS experiments are performed by mechanically deforming polymeric materials directly in the ion source housing of a time-of-flight mass spectrometer and mass analyzing the evolved volatile compounds. This technique has been applied to the study of stress-induced chemical reactions in polymeric materials, i.e., mechanochemistry, and to the characterization of residual volatile compounds in intractable polymer and composite matrices. Several polymeric systems ranging from polystyrene to fiber–epoxy composites have been studied by this technique. The significance of results achieved to date is assessed, and a systematic framework for further studies is developed.

The chemical analysis of polymeric materials is difficult and often time consuming. Usually a complete analysis, particularly of a new or unknown material, requires the use of several modern physical analytical instruments. The common methods use IR, visible, UV, and magnetic resonance spectroscopy; liquid and/or gas chromatography; differential thermal and thermogravimetric analysis; and mass spectrometry (*1*). In other cases, methods using changes in the mechanical properties of the material are used to monitor chemical changes; for example, tensile stress–strain or elongation can be used to monitor curing reactions (*1*). However, to characterize a polymer completely requires several analyti-

0-8412-0406-3/79/33-174-053$07.00/1

cal methods. Each method and associated instrumental technique has advantages and disadvantages. For example, IR measurements which reveal information about the functional groups in the polymer are usually performed on solid samples. Transmission IR spectroscopy is readily compatible with thin films, and Fourier transform IR (FTIR) spectroscopy can be used to monitor surface reactions. However, none of these methods is particularly useful in determining the chemical constituents of the polymer. Mass spectrometry, on the other hand, is an excellent method for studying the chemical composition of the polymer provided that the sample is sufficiently volatile. Several methods either volatilize the polymer or release the volatile fraction of the polymer into the ion source (*2*). Thermal methods—either direct pyrolysis inside the ion-source housing or a combination with gas chromatography, i.e., pyrolysis GC/MS—have been described previously as have other mass spectrometric techniques combined with different thermal analysis and thermogravimetric analysis (*3, 4*).

This chapter reviews another mass spectrometric method called stress mass spectrometry (stress MS). In stress MS, materials are subjected to mechanical deformation, and the volatile compounds evolved from the sample are analyzed by a mass spectrometer. The entire experiment, including application of stress, is performed directly in the ion-source housing of the mass spectrometer. Data from these experiments provide information on changes in the polymer which produce the evolved volatile compounds. When combined with the results of other spectroscopic experiments, these studies provide a means of investigating the chemical processes which occur when polymers are deformed mechanically.

Measurement Techniques

Stress MS has been applied to the study of polymers for 15 years, and several different devices have been developed for these studies. In the following sections, we discuss mass spectrometers for stress MS, devices for mechanically loading polymeric samples, and data acquisition and analysis techniques.

Mass Spectrometers for Stress MS. The requirements of a mass spectrometer for stress MS are more stringent than those for conventional mass spectrometry. The events to be monitored are short-lived, typically ≤ 1 sec, and the total amount of evolved compounds is small, between 10^{-9} and 10^{-10} g. In addition, a source vacuum housing versatile enough to accept readily the various devices for mechanically loading polymeric samples is required. The mass spectrometer must have a large, open ion source.

Table I. Sensitivities and/or Detection Limits of Stress MS Instrumentation

Reference	*Sensitivity/Detection Limit*
5	4×10^{-11} g/s - unit pk height
6	1×10^{-10} g/s - cm pk height
7	4×10^{-10} g/s - mm pk height
8	3×10^{-10} g/s - mm pk height[a]
9	4×10^{-9} g/s detection limit[a]
10	1×10^{-10} g/s detection limit
11	1×10^{-9} g/s detection limit

[a] Sensitivity calculated from data given in reference on the basis of compound with mol wt of 100.

A time-of-flight mass spectrometer (TOFMS) is an excellent instrument for these studies. The TOFMS produces approximately 10^4 spectra per sec. It is difficult to record individual mass spectra at this rate, but several data acquisition schemes (*see* below) have been developed which record a finite number of discrete spectra. These techniques follow changes in peak intensity over millisecond intervals, thus recording the mass spectra of compounds whose residence time in the ion source is short.

The sensitivities and/or detection limits of the stress MS system are summarized in Table I. The large, open ion source of the TOFMS together with its vacuum housing is ideally suited for insertion of the mechanical degradation apparatus.

Magnetic sector instruments generally are not suited for these studies. Although they have adequate sensitivity and greater resolution than the TOFMS, they scan too slowly, they have "tight" ion sources, and the source vacuum housing is not easily modified to accept the degradation subsystem. Quadrupole mass spectrometers could be used for stress MS studies, but no reference to their use in this application has been found. Thus, so far stress MS studies have been performed exclusively with time-of-flight mass spectrometers.

Techniques for Mechanically Deforming Polymer Samples. LOADING IN TENSION. The earliest work on stress MS (*12*) was performed by loading samples in tension. This method has been used frequently, and devices have been fabricated to load specimens cut from the bulk polymer (*6, 7, 9, 10, 12–16*), thin films (*12, 13, 14, 15*), fibers (*15*), or monofilaments (*11, 16*). A typical apparatus used to stress dogboned-shaped samples cut from sheets is shown in Figure 1. The sample is suspended from a lintel or other solid structure inside the ion-source vacuum housing, and the other end of the sample is attached to a loading rod by sample grips. The loading rod passes through a vacuum feedthrough, thus permitting the sample to be loaded mechanically by an apparatus outside the instrument vacuum. Frequently, the sample is loaded by adding a weight to a carriage attached to the loading rod. In some instances, the stress/strain history of the sample is measured by a strain gage (*9*) and/or extensometer (*9, 11*). The sample can be stressed with a fixed (*14, 15*), stepwise increasing (*6, 11, 14*) or continuously increasing force (*17*).

Regel' and Muinov (*6, 14*) used stepwise loading to determine the relationship between the applied stress and the evolution of volatiles from poly(methyl methacrylate) (PMMA) and polystyrene (PS). They also investigated the evolution of volatile compounds from these polymers subjected to a constant load (*13*).

For monofilament samples, the apparatus shown in Figure 2 is used to load the sample in tension (*11*). The monofilament is wound around a split cylindrical mandrel, half attached to the fixed lintel and half attached to the loading rod. The sample is attached to the mandrel at two points and is free to move around the mandrel as the sample is extended under load.

It is important to ensure that the data acquisition system is synchronized with the application of stress. This synchronization places severe limits on the type and method of data acquisition since the time available is several seconds. Therefore, it is necessary to ensure that sample failure

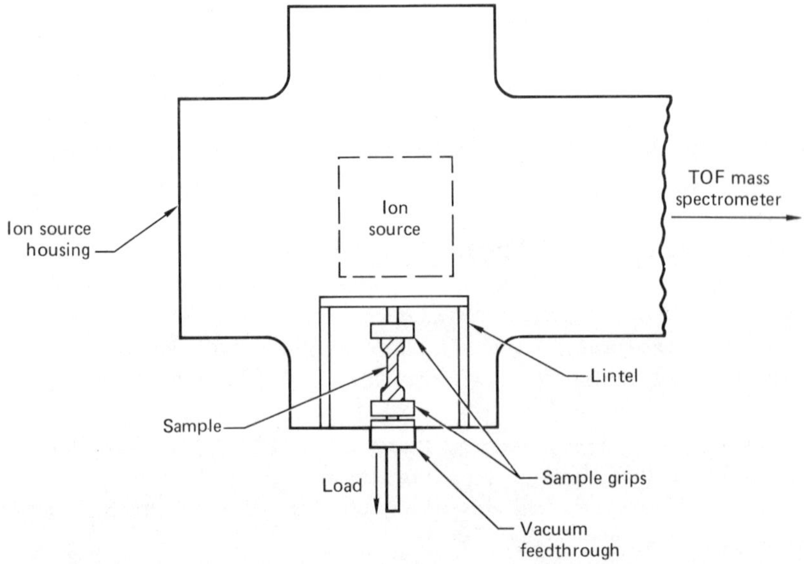

Figure 1. Typical apparatus for stress MS studies of bulk polymeric samples in tension

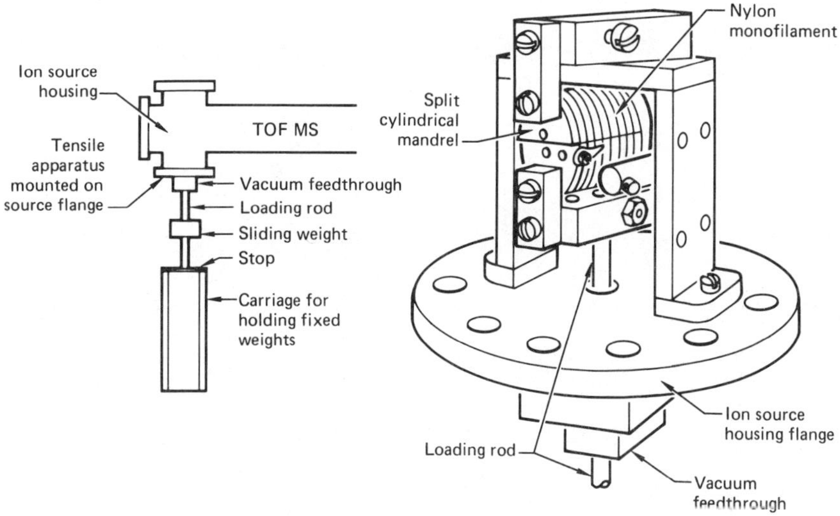

Figure 2. Apparatus for stress MS studies of monofilament samples in tension

occurs shortly after the load is applied. This can be done by scaling the sample cross section and loads such that failure occurs within seconds of loading (*12*).

An impulsive loading technique can be used to cause sample failure within tenths of seconds of the desired time. This method was proposed by Fanter (*18*) and was used to study the mechanical degradation of nylon 66 (*11*). A combination of fixed weights and impulsive loads is used to deform the sample. After the addition of each fixed weight, a mass concentric with the loading rod falls a fixed distance against a stop on the loading rod. If sample failure does not occur, the fixed weight is increased, and the impulsive load is reapplied. This procedure is continued until sample failure occurs.

Most stress MS work has been performed by loading samples in tension; however, a variety of other forms of mechanical deformation of the sample have been reported. These include bending to fracture, cutting, sawing, and crushing with a ball mill.

BENDING IN FRACTURE. This technique has the advantage of accommodating up to eight samples, and the time of catastrophic failure of the sample is easily controlled. However, this technique is useful only for obtaining mass spectra of the volatile compounds evolved from the sample and is thus limited to qualitative studies. The apparatus is shown in Figure 3. A turntable containing eight sample clamps arranged around the edge is used to hold samples with cross sections of several square millimeters. The turntable is rotated from outside the vacuum system to

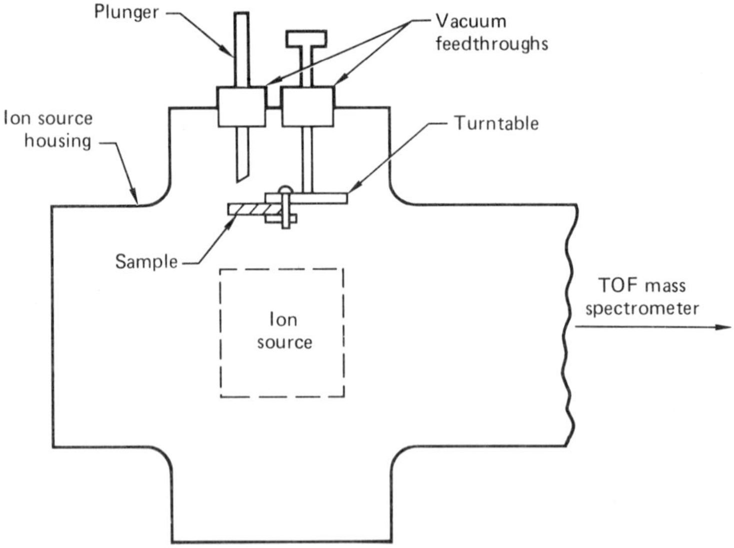

Figure 3. Apparatus for stress MS studies of bulk polymeric samples by bending

align the sample with the plunger. A window in the source flange is used to confirm alignment. Sample fracture occurs when the plunger is manually forced against the sample. A microswitch detects initial plunger movement and triggers the data acquisition system so that mass spectral data is acquired before, during, and after failure of the sample.

CUTTING. Mal'chevskii et al. used a cutting device to mechanically deform nitrocellulose (NC) (*7*). The time of failure is controlled precisely with this technique, but it too is limited to qualitative studies. Rotary motion of the cutting head, which has the appearance of an end mill, is accomplished by a special bellows vacuum feedthrough. When data from cutting were compared with those from loading in tension, changes in the relative height of peaks in the mass spectra were noted, but no explanation for these differences was given.

SAWING. Wolf and Grayson (*16*) used a slit saw to mechanically abrade composite samples. The advantages and limitations of this technique are similar to those for cutting. However, an appropriate data acquisition method (*see* continuous Ion Monitoring," below) and motor-driven rather than manual saw rotation provide both qualitative and quantitative studies of mechanical deformation in polymeric samples. A diagram of this device is shown in Figure 4. Vacuum feedthroughs permit rotation of the saw blade and sample feed to the saw. Although this device was intended primarily to study epoxy fiberglass composites, it could be used for other polymeric materials.

CRUSHING. Byl'skii et al. (8) used a vibration ball mill to study the volatile compounds released from a series of crushed polymers. They chose this method to intensify the mechanical degradation process and thus to increase the rate of volatile evolution from the polymer. For polymers such as polyethylene (PE) and polytetrafluoroethylene (PTFE), compound evolution rates are low from other methods of mechanical deformation; hence, it is difficult to obtain mass spectra. Byl'skii et al. used this technique to successfully obtain mass spectra of compounds from PE and PTFE and demonstrated the feasibility of the vibration ball mill for performing kinetic studies of mechanical degradation as a function of the amplitude of the vibration and duration of the grinding.

The apparatus consists of a flask containing balls which is connected by flexible bellows to the ion-source vacuum housing. The flask is vibrated by a motor drive outside the vacuum system. Volatile compounds evolved from the sample pass through the flexible bellows connection into the ion source.

Combination of flasks and balls of different materials were studied to determine which materials give the lowest outgassing rate when vibrated without a sample. They found that a stainless steel flask with molybdenum glass balls exhibits the lowest background.

An important feature of all these devices is that one or more vacuum feedthroughs are required for linear or rotary motion. A common problem with such feedthroughs is that small amounts of sorbed gases

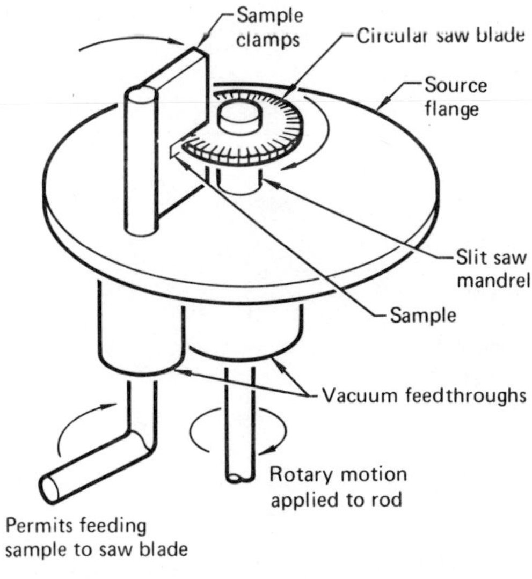

Figure 4. Apparatus for stress MS studies of bulk polymeric samples by sawing

may be evolved during motion of the feedthrough apparatus. These gases may be mistaken for volatile compounds evolved from the sample unless such artifacts are checked for. The bellows type tends to minimize these problems. However, these feedthroughs are not readily adaptable to rotary motion unless a sophisticated apparatus, such as that used by Mal'chevskii (*7*), is used. O-ring-sealed vacuum feedthroughs may be used provided that two such seals are used with differential pumping between them. This type of feedthrough is extremely useful since the same basic feedthrough can accommodate either linear or rotary motion. Further, such feedthroughs can be designed and fabricated easily from standard vacuum parts.

Data Acquisition Techniques

Before data acquisition techniques for stress MS experiments are reviewed, the operation of the TOFMS is discussed briefly. Detailed discussions of the theory and principles of TOFMS operation are in the literature (*19, 20*). The essential features of a TOFMS are shown in Figure 5. The elements A, B, C, E, and F provide a pulsed monoenergetic electron beam of fixed current which ionizes molecules in the ion-source

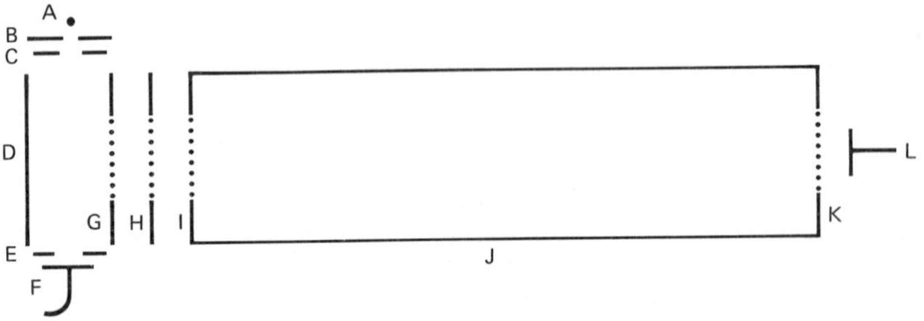

			Potential
A	Filament	Ion source	−50 to 70 Vdc
B	Electron control grid	Ion source	Pulsed
C	Ground grid	Ion source	Ground
D	Backing plate	Ion source	Ground
E	Ground grid	Ion source	Ground
F	Trap	Ion source	0 Vdc
G	Ion focus grid		Pulsed
H	Intermediate accelerating grid		−400 to −700 Vdc
I	Final accelerating grid		−2800 Vdc
J	Drift region		−2800 Vdc
K	Terminating grid for drift region		−2800 Vdc
L	Detector		

Figure 5. Schematic of basic elements in a TOFMS

region. The electron beam is pulsed by a suitable signal applied to the electron control grid (B). The bunched ions thus formed are extracted from the source by a negative pulse applied to the ion focus grid (G) and are thus accelerated by the two grids H and I which are at fixed negative potentials. All the ions in the bunch thus enter the field free-drift region with the same energy eV (where e is the charge on the ion and V is the final accelerating potential of grid I). Ions of lighter mass (m) travel through the drift region to the detector at a higher velocity than those of heavier mass; their velocity v is:

$$v = \sqrt{\frac{2eV}{m}}. \tag{1}$$

The time for an ion of mass 600 amu to traverse the drift region ($\sim$ 2 m) at an accelerating potential of 2800 Vdc is approximately 70 μsec. Thus, the entire mass spectrum from 1 to 600 amu can be analyzed repetitively 14,000 times per sec. Typical repetition rates for TOF mass spectrometers are approximately 10 kHz. The output signal from the electron multiplier is amplified and displayed on an oscilloscope which is triggered by the mass spectrometer repetition rate oscillator.

The essential data acquisition problem in stress MS experiments is to record the information displayed on the oscilloscope screen during the time that the sample undergoes stress and/or failure. Both cinematography and still photography of the oscilloscope screen have been used.

Cinematography. This data acquisition technique is straightforward in principle but tedious in practice. A high-speed movie camera photographs the mass spectral display on the oscilloscope screen. When the camera operates at 400 frames per sec and the mass spectrometer at a repetition rate of 10 kHz, each frame of the film contains the information of 25 mass spectra. This arrangement permits time resolution of 2.5 msec. A 150-m film records data for approximately 50 sec which is adequate for most stress MS experiments.

Cinematographic recording is the most frequently used data acquisition technique reported in the Russian work (*6, 7, 8, 13, 14*) and by Baumgartner et al. (*9*). This technique has the advantage of recording the heights of the various peaks as a function of time in the mass spectrum, thus providing qualitative data. Analysis of the data for peak heights, however, is tedious. Baumgartner et al. (*9*) report the automated analysis of cinematographic data by a flying spot scanner with subsequent digital data processing. The use of z-axis modulated mass spectral displays (*21, 22*) (*see* below) should enhance the cinematographic recording technique; however, none of the surveyed literature reports its use.

Still Photography. The use of a still camera to record the mass spectral data displayed on the oscillographic screen is an alternate photographic data acquisition technique. The simplest method is to photograph the oscilloscope display for a fixed time during which the sample is stressed and/or fails. This technique does not permit resolution of changes in the mass spectra with time but provides a way to record qualitative information with readily available equipment.

Time resolution of changes in the mass spectra can be accomplished in several ways. A TOFMS accessory is available from CVC Products, Inc. (Rochester, NY), which rasters mass spectra across the oscilloscope screen both vertically and horizontally. This device counts a preset number of spectra, *n*, and then offsets the trace both vertically and horizontally. After *n* more spectra, the trace is again offset. This procedure continues up to the desired number of traces. The device can be used to display 1–16 traces with 1–64 spectra per trace, thus permitting time resolution as low as 0.1 msec. However, the total display time, 16 traces times 64 spectra/trace, is only 0.1 sec. Thus, synchronization of sample loading and/or failure with this data acquisition device is critical. This data acquisition method provides both quantitative and qualitative data about the evolved compounds. However, the mass range which can be displayed effectively is limited to about 100 amu.

Another method for recording time-resolved changes in TOFMS with still photography was presented by Lincoln (*21, 22*). This technique uses *z*-axis modulation to intensify the oscilloscope trace during the individual mass spectral peaks. The oscilloscope intensity is adjusted so that the mass spectral baseline does not write on the oscilloscope screen; thus, the screen registers only when a peak is present. The mass spectrometer output signal is sent to the *z*-axis modulation circuit rather than the vertical input of the oscilloscope. A triggered ramp voltage either from a function generator or from a second oscilloscope is connected to the vertical input of the oscilloscope (Figure 6). In this data

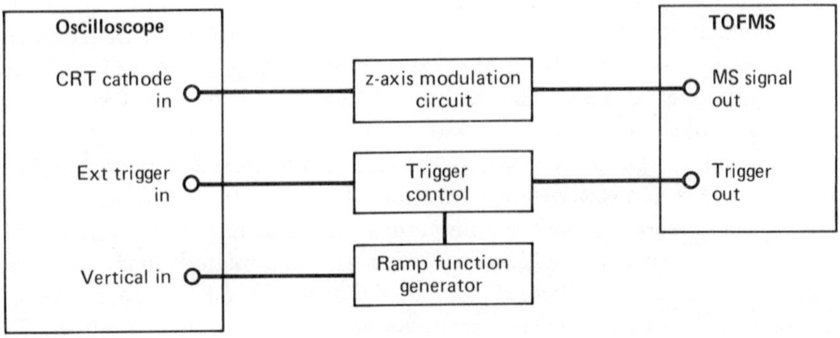

Figure 6. Block diagram of z-axis modulation system

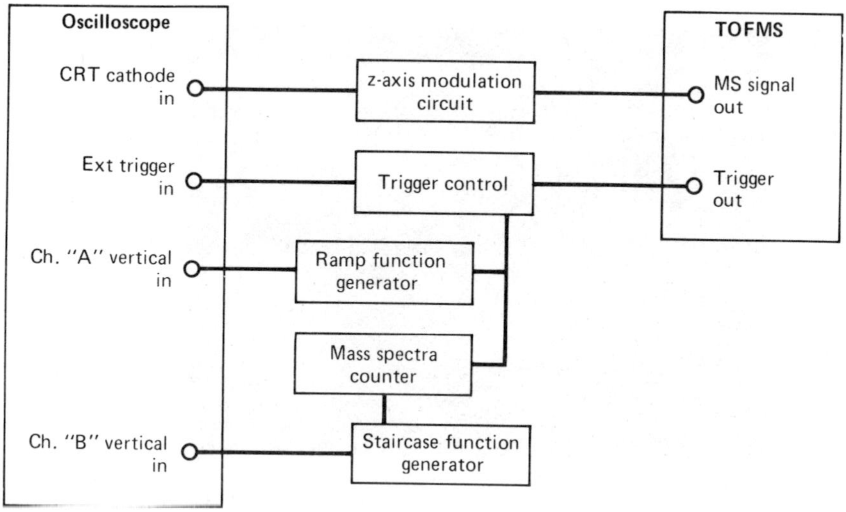

Figure 7. Block diagram of z-axis modulation system with internally generated timing lines

acquisition scheme elapsed time is measured along the vertical axis of the oscilloscope. The elapsed time can be adjusted according to the requirements of the experiment by either changing the slope of the ramp voltage or by changing the sensitivity of the vertical amplifier.

Grayson and co-workers (*10, 11, 16*) used a combination of z-axis modulation and mass spectral counting to record a data display with internally generated timing lines (Figure 7). The triggered ramp voltage is input to channel A of the vertical amplifier. However, channel B of the vertical amplifier is fed a staircase function obtained by counting a selected number of mass spectra. Channels A and B are summed with the vertical amplifier of the scope, and a display such as that shown in Figure 8 is obtained. The elapsed time of a vertical trace is the product of the number of spectra and the period of the TOFMS.

Although z-axis modulation techniques are useful for recording time-resolved mass spectra, they are not accurate enough to obtain quantitative data. In principle, the brightness of the trace on the photograph can be related to the intensity of the mass spectral peak. However, since the trace brightness is a function of many variables (threshold level of the z-axis modulation circuit, the brightness control of the oscilloscope, the photographic conditions used, and the type of film), it is difficult to quantitatively compare the results of different stress MS experiments. It is possible to compare the results from one trace on the film to the next by a densitometer. Nevertheless, accurate quantitative information can be obtained only after a calibration curve is derived for the photographic

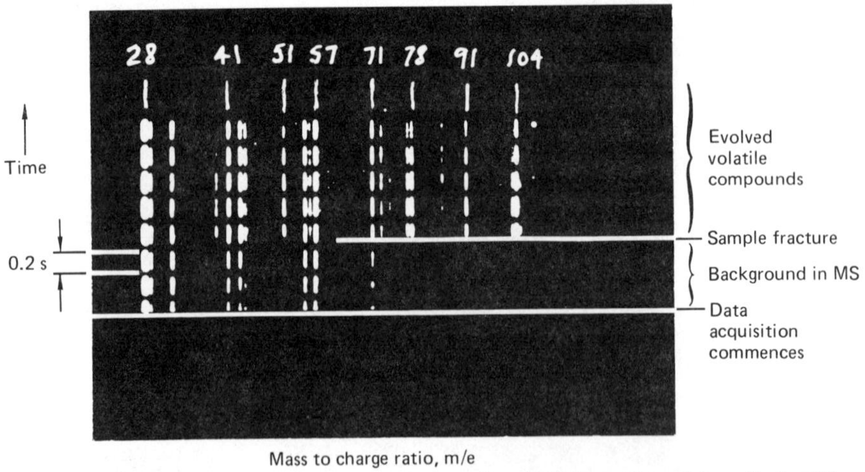

Journal of Polymer Science, Polymer Physics Edition

Figure 8. Photograph of z-axis modulated oscilloscope display obtained during fracture of PS (10)

emulsion by techniques similar to those used in mass spectography (*23*). This is not practical since the photographic properties of the film used for oscilloscope photography are not as well controlled as those of emulsions used in mass spectography.

A particularly important concern in all photographic data acquisition techniques, whether by cinematography or still photography, is choice of the phosphor for the cathode ray tube (CRT). It is desirable to choose a CRT whose light output has the highest actinic value. This requirement is particularly critical for recording data from a small number of spectra when the integrated light output from the CRT is small. Information concerning CRT phosphors and other general topics of oscilloscope photography are available from Tektronix, Inc. (*24*).

The photographic techniques discussed above are useful for obtaining a qualitative picture of the degradation process. Even the simplest photographic method can help to provide mass spectrometric data on the compounds evolved from a sample during a stress MS experiment. Thus, in almost every case, initial experiments should be performed using photographic techniques.

Analysis Techniques

Continuous Monitoring. Continuous ion monitoring is a mass spectrometric data acquisition technique with a long history which was recently reviewed by Falkner (*25*). Byl'skii (*8*) reports the utility of this technique for studying the kinetics of the mechanical degradation of

PMMA. Wolf and Grayson (*16*) used this technique to monitor the evolution of indigenous volatile compounds in composite materials and to study the mechanical degradation of PS (*10*).

In the continuous ion monitoring technique, the ion current of the ion of particular interest is recorded on a strip chart recorder throughout the experiment. The conventional analog scanner of a TOFMS can be used to monitor one ion continuously by setting the gate of the scanner at the mass of interest. A four-channel monitor available from CVC Products, Inc., permits monitoring of the ion current of four separate masses simultaneously. If desired, the total output integrator of the TOFMS can record the sum of the ion currents over a range of masses.

The principal advantage of this technique is that it provides accurate quantitative data on the ion current of a particular mass as a function of sample loading. The method has a good signal-to-noise ratio and a dynamic range greater than the photographic data acquisition techniques. Its major disadvantage is that it provides data from a limited number of ions during an experiment. In principle, the best TOFMS data acquisition system consists of continuous ion monitoring of all ions in the mass spectrum.

Low-Ionization-Potential Mass Spectrometry. Grayson et al. used low-ionization-potential mass spectrometry to study the mechanical degradation of products from PS (*10*) and nylon 66 (*11*). In this technique, the energy of the ionizing electron beam is lowered from 70 eV to 8–15 eV. The principal advantage is that mass fragmentation is suppressed; thus, the mass spectral peaks consist primarily of molecular ions. This approach is valuable in interpreting the mass spectrum of a mixture of compounds which usually occurs in stress MS experiments. It can be used to further advantage to identify isobaric mixtures. Grayson and Wolf (*11*) discuss the use and limitations of this technique.

The range of experimental methods and data acquisition techniques for stress mass spectrometry experiments provide a way to obtain a great amount of data on the volatile compounds evolved from polymeric samples when subjected to a mechanical load. These data provide information concerning the events occurring in the polymer which produce the evolved volatile compounds.

Studies of Stress-Induced Chemical Reactions

Kinetic Theory of Fracture. Catastrophic failure of a polymeric material is a complex process in which a sequence of partially understood events occurs at both the molecular and macroscopic levels. The stress-induced cleavage of the main-chain polymer bond is one event occurring on the molecular level which has been studied by both stress MS and electron spin resonance spectroscopy (ESR).

The first kinetic study of free-radical formation in mechanically loaded polymers was by Zhurkov and co-workers who studied the stepwise loading of nylon 6 and silk fibers (*26*). The concentration of free radicals was monitored by ESR as a function of applied stress and time. They found that the rate of radical formation (R) is a function only of stress (σ):

$$\frac{dR}{dt} = B \exp (\beta\sigma) \tag{2}$$

where B and β are constants characteristic of the material. Equation 2 was compared with the expression for the time-to-break t_b of a uniaxially stressed sample:

$$t_b = A \exp(-\alpha\sigma) \tag{3}$$

where again A and α are constants. According to Zhurkov, when and if the rate of radical formation is the decisive factor determining the lifetime of a stressed polymer, the coefficients α and β are equal, and the product $t_b(dR/dt)$ should be a constant. Their experimental data confirmed this prediction. Previously, Tobolsky and Eyring had proposed that the breaking time of polymeric threads was an exponential function of stress (*27*). This led Zhurkov to propose a modified Arrhenius relation which describes the time-to-failure as a function of energy, stress, and temperature:

$$t_b = t_o \exp[(U_o - \gamma\sigma)/kT] \tag{4}$$

where t_b is the rate of bond rupture in the polymer chain, t_o is a constant related to fundamental frequencies of the solid, U_o is the activation energy for bond rupture, γ is a structural coefficient which defines the actual loads in a stressed body, σ is the applied stress, k is the Boltzmann constant, and T is the temperature (K). At zero applied stress, Equation 4 reduces to the classical Arrhenius equation relating the time-to-failure to the temperature. According to the kinetic theory of fracture, a close correlation should exist between mechanical degradation and thermal degradation. The elementary events in both processes are the cleavage of chemical bonds in the polymer chain accompanied by the formation of free radicals on the ruptured ends of the chain. The primary free radicals react and/or rearrange rapidly to form more stable radicals. These reactions may generate products which evolve during mechanical degradation. The detection and analysis of trace quantities of low-molecular-weight compounds which may be evolved upon fracture are suited to mass spectrometry provided that the compounds can be introduced to the ion source.

Stress MS Studies. The first report on the use of mass spectrometry to investigate the mechanical degradation of polymers was published in 1962 by Regel et al. (*12*). The samples were fractured directly in the ion source housing of a TOFMS. They studied the volatile products from PMMA and PS to try to corroborate Zhurkov's theory. They concluded that a definite correlation exists between mechanical and thermal degradation but noted that these results pertain only to polymers in which thermal degradation is initiated by main-chain rupture and that thermal degradation usually begins at 300°–400°C, temperatures which greatly exceed those used in mechanical studies. They, as well as the subsequent workers in this field, did not determine the residual monomer remaining in the matrix before fracture. The samples were vacuum-annealed for several days at temperatures above the glass transition temperature (T_g) to remove residual volatile compounds. Regel et al. (*6*) discussed their studies in greater detail and concluded that both mechanical breakdown and thermal degradation are based on elementary reactions following rupture of chemical bonds.

Regel' and Muinov investigated the volatile products released from PMMA, PS, and polyvinyl alcohol (PVA) as a function of load (*13*). They reported that volatile products were released from the instant that the load was applied with a sharp increase at fracture. The curve of monomer evolution as a function of stress had the same shape as a creep curve; an observation they interpret as direct proof that the chain rupture process begins the moment the load is applied. Their system could detect as little as 10^{11} molecules/sec, which according to the authors is several orders of magnitude more sensitive than ESR. In addition to noting that breakdown occurs at the instant the load is applied, they studied the stepwise addition of a load. From both PMMA and PS, they observed an abrupt increase in monomer, measured as the height of the parent molecular ion, which quickly reached a steady-state value (h_{st}). The volatile concentration was an exponential function of stress, i.e.:

$$h_{st} \; \alpha \; \exp(\text{stress}), \qquad (5)$$

an observation in excellent agreement with the theoretical predictions of Zhurkov (*see* Equation 2).

Later Pozdnyakov and Regel' noted that three separated regions appear in the height of the monomer peak (h_{104}) from stressed PS (*17*). Only in the high stress region, i.e., region III, is the rate of monomer release proportional to applied stress. In the low stress regions(regions I and II), the monomer observed was a direct result of surface cracking. The volatile products observed in these studies are attributed to bond cleavage and to their associated secondary radical reactions.

Amelin et al. (*15*) compared thermal degradation and mechanical breakdown products of polymers by incorporating a stress-strain apparatus and a pyrolyzer into the same TOF. In addition, they investigated polymers such as polyvinyl chloride (PVC) and polyacrylonitrile (PAN), whose thermal degradation is initiated by fragmentation of the side-chain groups rather than main-chain scission. They conclude that for polymers such as PPMA, PS, and polypropylene (PP), whose mechanical and thermal degradation begin with the rupture of the main-chain, the mass spectra of the degradation products are essentially identical. However, for polymers such as PVC and PAN the mass spectra of the mechanical and thermal degradation products are different. They attribute this difference to the fact that mechanical degradation is initiated by rupture of the bonds in the main chain while the thermal degradation begins with the stripping of the side groups. Again, these results agree with the Zhurkov theory.

Mal'chevskii et al. (*7*) studied NC which, according to earlier reports (*28*), exhibited different mechanical and thermal initiation reactions. They concluded that the initial elementary reactions in the thermal and mechanical degradation of NC are dissimilar. The thermal degradation of NC begins with the rupture of the $RO–NO_2$ bond, but mechanical destruction begins either with the opening of 1–4 glucoside bonds or with the opening of some bond of the glucopyranose unit. The product distribution from the two forms of degradation was different; however, the activation energies for both processes were, surprisingly, the same, i.e., about 159 kJ/mol (38 kcal/mol).

Byl'skii et al. (*8*) attached a vibration ball mill directly to the vacuum chamber of a TOF and studied the volatile products released during the vibration grinding of PMMA, PTFE, PE, and poly-α-methylstyrene (PMS). Considerable difference exists in vibration grinding and direct mechanical fracture. In the case of grinding, one expects a local heating effect attributable to the action of the grinding balls. Volatile products different from those in the background gases were observed as soon as the vibration commenced and decreased rapidly when the vibration stopped. They determined the rate of volatile product (usually monomer) formation.

Baumgartner et al. (*9*) compared the thermal and mechanical degradation of filled and unfilled elastomers. They were particularly interested in the long-term aging and fatigue behavior of solid propellants filled with ammonium perchlorate or potassium chloride. They reported that the mechanisms for thermally and mechanically induced decomposition of the propellant binder appear equivalent. At low temperatures, mechanical processes control the decomposition rates of the polymers, whereas thermal processes control the decomposition at high temperature. They further note that, "Equivalence of thermal and mechanical degradation mechanisms allows use of rapid pyrolysis techniques to study the molecular scale factors that control mechanical failure in polymers and composites."

Recently, Grayson et al. (*10*) investigated the mechanical degradation of PS. They used special procedures to purify the polymer from monomer and found that prolonged high-temperature vacuum annealing was not sufficient to remove monomer. Styrene evolved from both as-received and vacuum-outgassed samples; however, essentially no styrene was observed from fractionally reprecipitated PS. Contrary to the work previously reported by Regel' and co-workers, they report that the amount of styrene released during the mechanical fracture of carefully purified PS is small and that the number of primary chains broken is less than 10^{10} mm^{-2}.

Grayson and Wolf (*11*) also investigated nylon 66, a polymer whose residual impurities and anticipated degradation products are different. They modified the ion source housing of their TOF so that both mechanical and thermal degradation could be investigated by the same analytical detection scheme. They note that general agreement exists between the products observed in mechanical and thermal degradation. In both cases,

Table II. Reviewed Stress MS Literature

	Mechanical Deformation by				
Polymer	*Tension*	*Bending*	*Cutting*	*Sawing*	*Crushing*
Polymethylmethacrylate	6, 8, 12–15, 17, 29				
Polystyrene	6, 10, 12–15, 29	10			
Polyvinyl alcohol	13				
Polypropylene	15				
Polyacrylonitrile	15				
Polyvinyl chloride	15				
Nitrocellulose			7		
Polyethylene					8
Polytetrafluorethylene					8
Poly α methyl styrene					8
Hydroxy-terminated polybutadiene filled with ammonium perchlorate	9				
Hydroxy-terminated polybutadiene filled with potassium chloride	9				
Nylon 66	11, 16				
Epoxy fiberglass composite				16	

cyclopentanone, ammonia, carbon dioxide, and water are observed. However, major differences, particularly in the product distribution, exist.

For comparison and reference, stress MS studies on different polymers with different forms of mechanical deformation using different mechanical degradation techniques are summarized in Table II. The major portion of the work has been done on relatively simple polymers, PMMA and PS, by loading in tension. Most of the work was performed by scientists at the A. F. Ioffe Physicotechnical Institute of the Russian Academy of Sciences in Leningrad.

Studies of Indigenous Volatiles

Although most stress MS studies to date have focused on stress-induced chemical reactions in polymeric materials, this analytical technique has proven utility in another important area of polymer research, namely characterization of volatile compounds indigenous to the polymeric matrix. It is possible that applications in this area could overwhelm mechanochemical applications.

Many analytical methods for characterizing indigenous volatile compounds are described in the literature. Gas chromatography has been used to analyze polymers for plasticizer content (*30*), residual monomer (*1, 31*), and nonpolymerizing impurities (*10, 11, 32*). Grayson and coworkers (*10, 11*) use a desorption technique which is a variation of the systems described by Levy et al. (*33*) and Ligon et al. (*34*).

A significant disadvantage of most of these methods is that the sample must be heated; consequently, thermally labile compounds may be destroyed, and tightly bound (or high-molecular-weight) compounds may not be released. Stress MS is performed at or near room temperature; thus, thermally labile volatile compounds are not destroyed, and the virgin surface in the high vacuum may release otherwise tightly bound compounds. Grayson and Wolf (*35*) report the detection of chlorophenylisocyanate in an epoxy–fiberglass composite by stress MS. This compound was not detected when the indigenous volatiles were characterized by thermal desorption GC/MS analysis.

Stress MS can be used also to determine quantitatively the distribution of a particular indigenous compound as a function of position in the sample. Wolf and Grayson (*16*) demonstrated that volatile compounds evolved during abrasion of the sample can be quantitatively measured by continuously recording the ion current of an ion characteristic of the compound of interest. The molecular ion of toluene ($m/e = 92$) was recorded continuously during abrasion of an epoxy–fiberglass composite. The ion current as a function of sawing is shown in Figure 9. Artifacts were checked for by rotating the saw freely and by sawing the aluminum clamp used to hold the sample. Toluene is detected only during abrasion

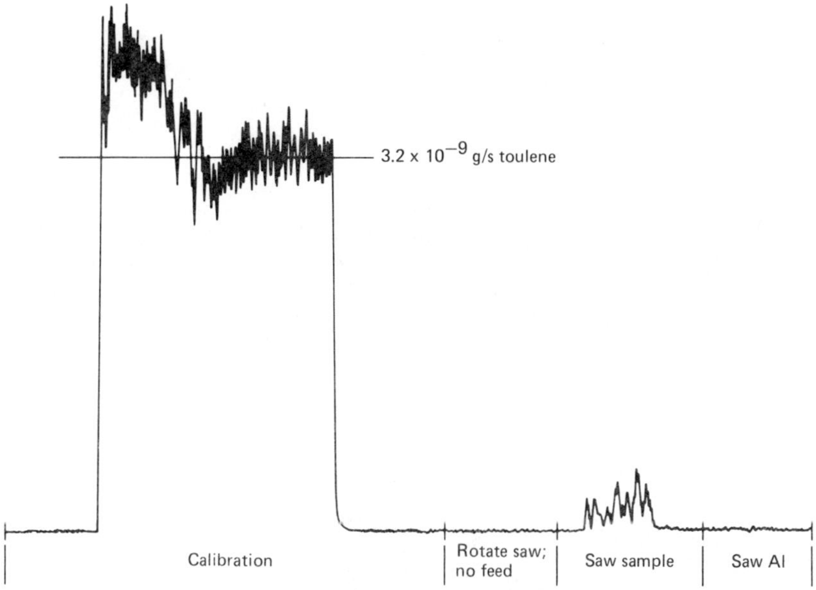

Figure 9. Ion current of m/e *92 (toluene) as a function of sample abrasion*

of the sample. The jagged variations in the ion current result from the manual rotation of the saw.

A precision abrasion apparatus (Figure 10) which will be used to determine quantitatively the distribution of indigenous volatile compounds in polymeric materials is under construction. A precisely known volume of the sample will be abraded by a motor-driven tool and a motor-driven three-degree-of-freedom sample stage. Continuous ion monitoring of the evolved compounds will permit quantitative analysis of their distribution in the polymer sample.

It is apparent from these studies that stress MS of polymeric and composite materials is a practical method to analyze for compounds trapped within the matrix. The technique is not meant to replace existing thermal desoprtion GC/MS techniques but is complementary to them. The primary advantages of stress MS as a way to characterize indigenous volatile compounds are that labile compounds can be detected and that information on the distribution of the compounds in the sample can be obtained. However, this method is not as sensitive as conventional methods in which the evolved volatile compounds are concentrated prior to analysis. Furthermore, it is necessary to interpret the mass spectrum of the mixture of compounds. Nevertheless, the difficulties associated with the determination of impurities of unreacted compounds in an intractable polymeric matrix warrants the continued development of stress MS.

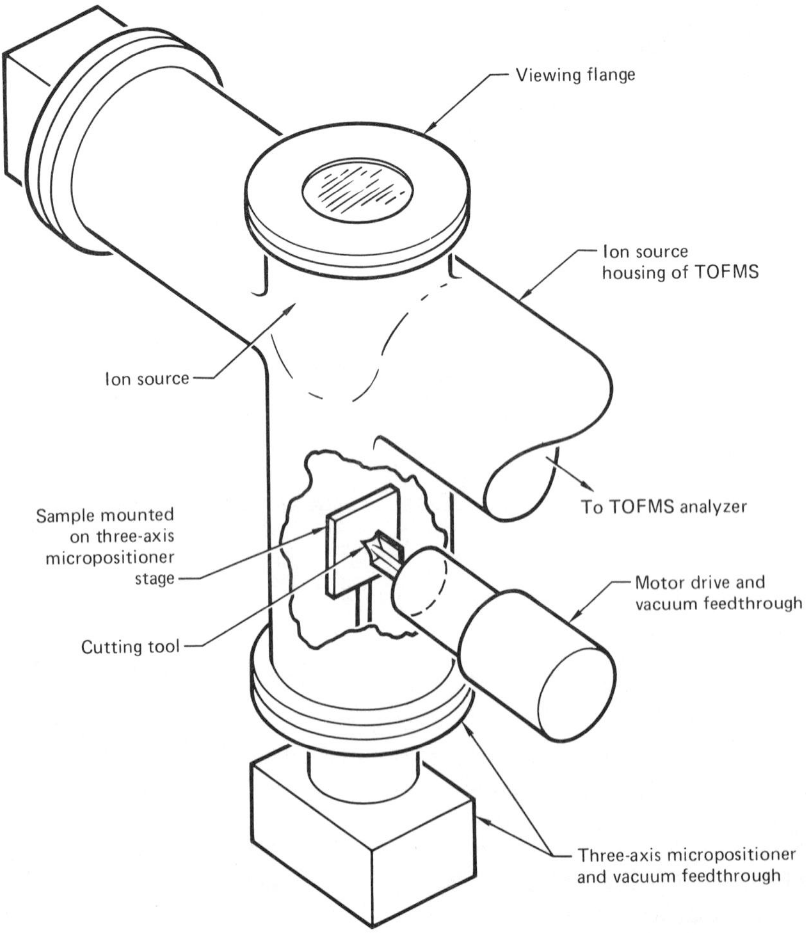

Figure 10. Precision abrasion apparatus for quantitatively determining the distribution of volatile compounds indigenous to a polymeric matrix

Discussion

The major thrust of the stress MS studies to date has focused on gathering evidence to support the kinetic theory of fracture proposed by Zhurkov. For the Zhurkov kinetic theory of fracture to be applicable to the material in question, not only must the mechanical and thermal degradation products be identical, but both processes must follow the same degradation mechanism.

The major questions which must be addressed to develop a mechanistic picture of mechanical degradation are summarized in Figure 11. Specifically, we must answer the following questions:

(1) Are volatile products released upon fracture?
(2) Are they degradation products or indigenous compounds?
(3) Are main-chain bonds broken?
(4) Are the thermal and mechanical products the same?
(5) Are the kinetics the same?

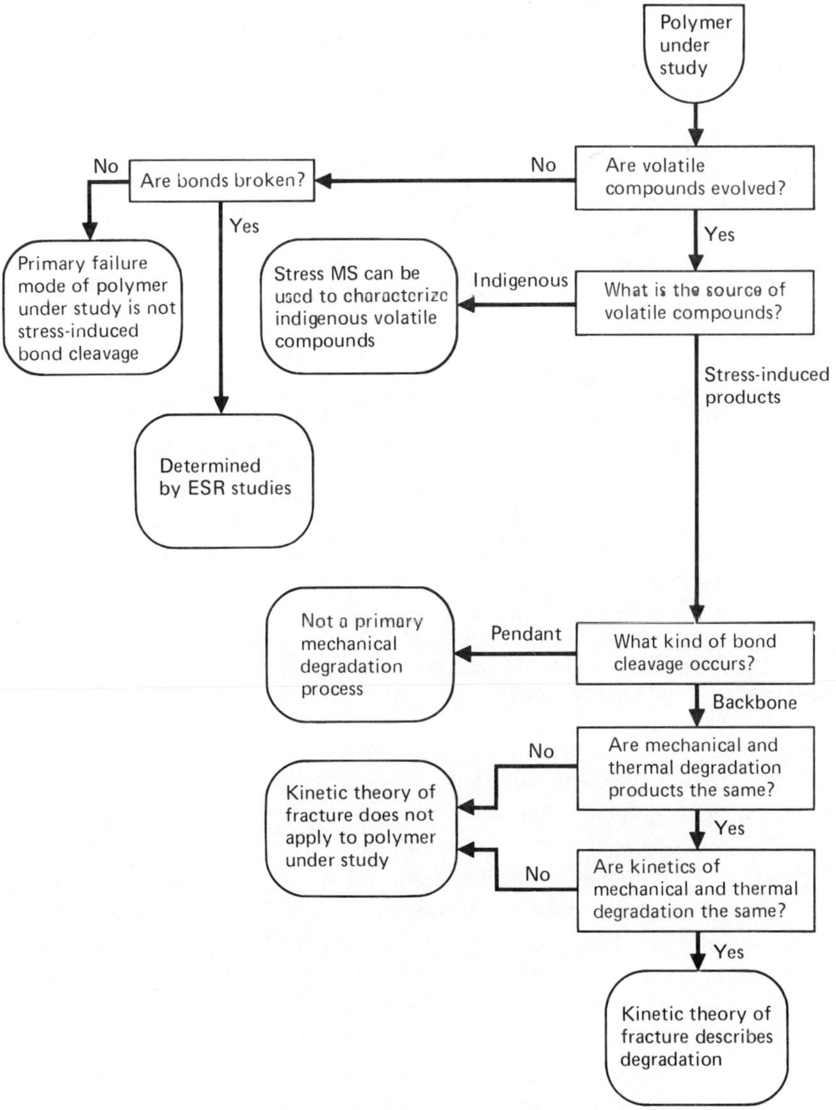

Figure 11. Important questions in the use of stress MS data as evidence for the kinetic theory of fracture

Only if the answers to all these questions are affirmative can we use the Zhurkov kinetic theory of fracture to explain the degradation processes. The evidence gathered to date which is germane to this subject is summarized below.

Most investigators observed volatile compounds released when the polymeric material was subjected to mechanical forces. However, the source of the volatile compounds is questionable; we must determine whether the observed products were indigenous to the matrix or are formed by the degradation process. This problem was recognized clearly by the early workers in the field. Regel' et al. (*12*) compared the products released from vacuum-degassed and vacuum-baked samples of different thicknesses. Regel' and Muinov address this problem in their study of PMMA, PS, and PVA but only mention that, "special experiments [are required] to prove the volatile products liberated on rupture were not contained in the specimen beforehand" (*13*). However, they did not describe the "special experiments." In another publication on PMMA and PS by these same authors (*14*) the problem is discussed again, and they note:

"To avoid this effect [artifacts from indigenous monomer] it is necessary either to use test specimens prepared from reprecipitated polymers, which are free from the monomer and other low-molecular-weight fractions, or to take special precautions to remove volatile products by heating the specimen in vacuum. For this purpose in our experiments, the specimens were heated before test in a 10^{-6} Torr vacuum (up to 150°C), sometimes for several days. The experiments on specimen breakdown were carried out at ~ 40°C after the mass spectrometer, at its maximum sensitivity, had ceased to register any yield of free monomer or other volatiles from the specimen. It should, however, be appreciated that such treatment did not guarantee that absolutely no volatile substances were left, particularly in thick specimens at some distance from the surfaces."

They clearly recognize the problem and even suggest two potential solutions. However, they used the wrong purification method to ensure monomer-free polymer (*see* Ref. 10.)

A brief reference to the problems of indigenous volatile compounds was made by Regel' et al. on the stress MS of PMMA and PS (*6*). Later, Regel' and co-workers (*15*) discuss this problem when comparing the products of mechanical and thermal degradation of PMMA, PS, PP, PAN, and PVC. They point out that in their earlier work (*12*) it was not possible to differentiate between the monomer existing in the free state of PMMA and the monomer arising from mechanical degradation. In their later work, they attempt to use the ratio of the ion currents at mass 59 (which is associated with an impurity in the polymer) and 100 (the parent ion of the monomer) to differentiate between indigenous monomer and mechanical degradation products. By rather circuitous reasoning, they state that the ratio of the 100 to 59 peaks can be used to differentiate "between the monomer fraction formed as a result of degradation and the fraction diffusing from inside the specimen" (*15*). Pozdnyakov and Regel' (*17*) conclude that compounds evolved during stress MS experiments of PS result from mechanical degradation as well as from free volatile products already in the polymer. Baumgartner et al. (*9*) do not discuss the problem but point out that the samples to be tested were "inserted into the mass spectrometer and pumped on until the outgassing reached a low level" prior to the application of stress. While all the above-mentioned authors recognized the extreme importance of this point, no studies were conducted to determine the amount of residual monomer remaining in the matrix.

Grayson et al. (*10*) investigated untreated, vacuum-baked at 120°C, and fractionally reprecipitated PS. They analyzed the various samples for indigenous volatile content by vaporization–gas chromatography/mass spectrometry (*33, 36, 37*) and found that fractional reprecipitation was the only effective method to remove the volatile fraction from PS. Further, stress MS experiments with PS samples prepared from the fractionally reprecipitated polymer evolved only a trace of styrene monomer.

Therefore, considerable ambiguity exists concerning the origin of the volatile compounds reported by the Russian workers. The only study in which residual monomer was shown to be absent (*10*) did not observe mechanical degradation products even though a sensitive TOF was used. Condensation-type polymers appear to be particularly attractive for further study because the expected degradation products and the starting materials (and associated impurities) are different; hence one can readily distinguish between products and trapped residual compounds.

Nylon 66 is a polymer whose residual impurities and degradation products are different. Grayson and Wolf (*11*) preceded their study on the thermal and mechanical degradation of nylon 66 by a careful analysis of the indigenous volatile compounds. A general agreement between the mechanical and thermal degradation products was observed (*11*). In both cases, cyclopentanone, ammonia, carbon dioxide, and water were observed. However, major differences in the product distribution were noted. For example, the most pronounced thermal degradation product is cyclopentanone. While this compound is formed during mechanical degradation, many other hydrocarbon ions including ethylene are observed in the spectra. The ammonia-to-cyclopentanone ratio is larger in the mechanical degradation experiments.

When the products result from the application of mechanical forces, the next question is concerned with the location and type of chemical bonds broken to produce the products. A wealth of data from ESR studies of many polymeric materials indicates that, at least at temperatures well below T_t, the polymer backbone, i.e., main-chain, ruptures at failure (*38, 39*). The failure mechanism associated with side- or pendant-group cleavage is not a primary mode of mechanical degradation although it is a well known effect in the thermal degradation of some polymers (*40*).

Amelin et al. (*5*) concluded correctly that the differences observed in the thermal and mechanical degradation of PAN and PVC result from pendant and main-chain cleavage, respectively. Mal'chevskii et al. report differences in the distribution of products from thermal and mechanical degradation of NC. Baumgartner et al. (*9*) conclude from their studies on filled hydroxy-terminated polybutadiene binders that the mechanisms of thermally and mechanically induced decomposition appear to be equivalent. Grayson and Wolf (*11*) conclude that some correlation exists between mechanical and thermal degradation products of nylon 66, but the correspondence is not one-to-one, and different mechanisms are probably involved. Amelin et al. (*15*) observed differences in the products of thermal and mechanical degradation of PAN and PVC, whereas the products from PMMA, PS, and PP were similar. Once again, because the nature of the indigenous compounds is not known, the results of PMMA and PS studies cannot be used to support the hypothesis that thermal and mechanical degradation mechanisms are the same.

Zhurkov and Tomashevskii (*41*) using time-to-fail studies calculated the activation energy U_0 for macroscopic failure and compared these values with those obtained thermally, i.e., E_0 for the degradation of several polymers. These results are summarized in Table III. They assume that the degradation processes in mechanical and thermal breakdown are identical because the activation energies are identical. These activation energies are considerably less than the dissociation energy of a typical C–C bond. If all these observations are reliable and if the measured activation energies are indicative of elementary processes, the overall degradation probably proceeds via a chain mechanism. This is an important point because Zhurkov later concluded that each primary

Table III. Activation Energies for Mechanical (U_0) and Thermal (E_0) Degradation (*41*)

Polymer	U_0 *(kJ/mol)*	E_0 *(kJ/mol)*
PVC	146	134
PS	226	230
PMMA	226	218–222
PP	234	230–243
PTFE	314	318–335
Nylon 6	188	180

Physical Basis of Yield and Fracture

Table IV. Activation Energies for Mechanical (U_o) and Thermal (E) Degradation (*43*)

Polymer	U_o (*kJ/mol*)	E_W (*kJ/mol*)	E_{MS_1} (*kJ/mol*)	E_{MS_2} (*kJ/mol*)
PMMA	130	218	126	209
PS	138	230	146	230
PE	105	264	84–105	293

Soviet Physics—Solid State

radical generated a vast number of ruptured polymer chains as a result of mechanical stress (*42*). He estimated that each primary bond broken produced as many as 10^3 stable molecules.

Amelin et al. (*43*) also compared activation energies for the thermal and mechanical destruction of polymers. The thermal activation energies were measured from weight loss (i.e., thermogravimetric analysis) and from the volatile product yield (measured in a mass spectrometer). Their results are summarized in Table IV. The two values reported for the thermal mass spectrometric studies were derived from best fits to the reaction rate/inverse temperature plots of the data. They conclude that the activation energy for mechanical degradation U_o and the activation energy for the initial stages of thermal degradation E_{MS_1} are equal. According to these authors, this correlation results from weak bonds in the polymer which control the mechanical properties.

This correlation between the activation energies observed in thermal and mechanical degradation has led many investigators to believe that their stress MS results correlate well with thermal data, however, the essential feature of stress MS requires that the observed compounds be degradation products. The mechanical activation energies reported by Zhurkov and Amelin for the degradation of PMMA and PS, differed considerably, yet both groups used time-to-fail to determine these values.

While these results certainly indicate the utility of stress MS for studying the mechanical degradation of polymers, they provide no positive support for the Zhurkov theory. Those studies which report a similarity in thermal and mechanical degradation products are suspect because the origin of the observed products is not known. Those studies in which degradation products are clearly distinguished from residual impurities, i.e., nylon 66, PAN, and PVC, apparently have different thermal and mechanical degradation mechanisms.

Summary

The complete analysis of a polymeric material is a difficult and complex problem. Stress MS studies have great potential to help solve one of the major industrial problems concerning the use of polymers—

their long-term durability. The usefulness of stress MS to relate thermal reactions to mechanical time-to-break or fail has been studied for several years. In the two most extensively studied materials, PS and PMMA, it is difficult to distinguish between degradation products formed by reactions of the primary cleavage radicals and the residual monomer which remains after polymerization. Most of the investigators erroneously assume that a vacuum anneal near the polymer T_g removes all the volatile impurities remaining after polymerization and processing. The single in-depth study concerned with this particular problem, i.e., removal of indigenous volatiles, showed that only fractional reprecipitation from solution yielded a polymer sample sufficiently pure for meaningful stress MS studies, at least for PS. Nylon, another material of great interest, is free of this problem because the expected degradation products are different from the residual volatile impurities. The mechanical and thermal breakdown paths, however, are different. The mechanical degradation of nylon proceeds via main-chain scission, and the thermal degradation results from the hydrolysis of the amide bond followed by secondary reactions (*44*). In other materials, such as NC, the activation energies for thermal and mechanical degradation are identical, yet the products released, and hence the mechanisms, are different.

The Zhurkov theory relating temperature and applied stress to time-to-fail represents a neat and concise picture of mechanical failure. However, a great deal of the experimental evidence gathered to date to support this theory is suspect. Additional studies are needed to either confirm or to deny this simple mechanistic picture of mechanical failure.

Although stress MS and the entire concept of a thermal–mechanical stress correlation may be in error, stress MS studies may have an even greater potential to help characterize volatile compounds trapped in or indigenous to otherwise intractable materials such as composites.

Acknowledgment

The authors are grateful to Roger J. Morgan for helpful and stimulating discussions.

Literature Cited

1. Cobler, J. G., Chow, C. D., *Anal. Chem.* (1977) **49**, 159R.
2. Hummel, D. O., Schueddemage, H. D., Ruebenacker, K., *Monogr. Mod. Chem.* (1974) **6**, 355.
3. Mol, G. J., Gritter, R. J., Adams, G. E., *Polym. Prepr., Am. Chem. Soc., Div. Polym. Chem.* (1976) **17**(2), 758.
4. McFadden, W. H., "Techniques of Combined Gas Chromatography/Mass Spectrometry: Applications in Organic Analysis," p. 157, John Wiley, New York, 1973.

5. Anufriev, G. S., Pozdnyakov, O. F., Regel', V. R., *Vysokomol. Soedin.* (1966) **8**(5), 834. [*Polym. Sci. USSR (Engl. Transl.)* (1966) **8**(5), 916.]
6. Regel', V. R., Muinov, T. M., Pozdnyakov, O. F., "Conference Proceedings: Physical Basis of Yield and Fracture," p. 194, Oxford, 1966.
7. Mal'chevskii, V. A., Pozdnyakov, O. F., Regel', V. R., Fal'kovskii, M. G., *Vysokomol. Soedin.* (1971) **A13**(9), 2078. [*Polym. Sci. USSR. (Engl. Transl.)* (1971) **13**, 2334].
8. Byl'skii, B. Ya., Pozdnyakov, O. F., Regel', V. R., Redkov, B. P., *Mekh. Polim.* (1973) **9**(5), 835. [*Polym. Mech. (Engl. Transl.)* (1973) **9**(5), 737].
9. Baumgartner, W. E., Hammond, J. A., Myers, G. E., Stapleton, W. G., *Polym. Prepr., Am. Chem. Soc., Div. Polym. Chem.* (1973) **14**(1), 448.
10. Grayson, M. A., Wolf, C. J., Levy, R. L., Miller, D. B., *J. Polym. Sci., Polym. Phys. Ed.* (1976) **14**, 1601.
11. Grayson, M. A., Wolf, C. J., "Applications of Polymer Spectroscopy," E. G. Brame, Jr., Ed., Academic, New York, p. 221.
12. Regel', V. R., Muinov, T. M., Pozdnyakov, O. F., *Fiz. Tverd. Tela* (1962) **4**(9), 2468. [*Sov. Phys.-Solid State (Engl. Transl.)* (1963) **4**(9), 1809].
13. Regel', V. R., Muinov, T. M., *Vysokomol. Soedin.* (1966) **8**(5), 841. [*Polym. Sci. USSR (Engl. Transl.)* (1966) **8**(5), 923].
14. Regel', V. R., Muinov, T. M., *Fiz. Tverd. Tela* (1966) **8**(8), 2364. [*Sov. Phys.-Solid State (Engl. Transl.)* (1967) **8**(8), 1884].
15. Amelin, A. V., Muinov, T. M., Pozdnyakov, O. F., Regel', V. R., *Mekh. Polim.* (1967) **3**(1), 80. [*Polym. Mech. (Engl. Transl.)* (1967) **3**(1), 54].
16. Wolf, C. J., Grayson, M. A., "Proceedings of the TTCP3 Critical Review: Techniques for the Characterization of Polymeric Materials," p. 117, Army Materials and Mechanics Research Center, Watertown, Mass., 1977.
17. Pozdnyakov, O. F., Regel', V. R., *Fiz. Tverd. Tela* (1968) **10**(12), 3664. [*Sov. Phys.-Solid State (Engl. Transl.)* (1969) **10**(12), 2905].
18. Grayson, M. A., Levy, R. L., Wolf, C. J., Fanter, D. L., *Polym. Prepr., Am. Chem. Soc., Div. Polym. Chem.* (1976) **17**(2), 764.
19. Wiley, W. C., McLaren, U. H., *Rev. Sci. Instrum.* (1955) **26**(12), 1150.
20. Agishev, E. I., Ionov, N. I., *Zh. Tekh. Fiz.* (1958) **28**, 1775. [*Sov. Phys.-Tech. Phys. (Engl. Transl.)* (1958) **3**(8), 1638].
21. Lincoln, K. A., *Rev. Sci. Instrum.* (1964) **35**(12), 1688.
22. Lincoln, K. A., *Int. J. Mass Spectrom. Ion Phys.* (1969) **2**, 75.
23. Beynon, J. H., "Mass Spectrometry and Its Application to Organic Chemistry," p. 60, Elsevier, New York, 1960.
24. Marsh, W., "Oscilloscope Camera Concepts," p. 107, Tektronix, Beaverton, Oregon, 1973.
25. Falkner, F. C., *Biomed. Mass Spectrom.* (1977) **4**(1), 66.
26. Zhurkov, S. N., Savastin, A. Ya., Tomashevskii, E. E., *Sov. Phys.–Dokl. (Engl. Transl.)* (1964) **9**, 986.
27. Tobolsky, A., Eyring, H., *J. Chem. Phys.* (1943) **11**, 125.
28. Mal'chevskii, V. A., Regel', V. R., Fal'kovskii, M. G., *Vysokomol. Soedin.* (1971) **B13**, 119.
29. Pozdnyakov, O. F., Regel', V. R., *Mekhanoemissiya Makhanokhim. Tverd. Tela* [*Dokl. Vses. Simp.*] 2nd 1969 (Pub. 1974), 207.
30. Sakata, M., Higasikuze, H., Kishne, K., *Shimadzu Hyoron* (1970) **27**(4), 281.
31. Li Gotti, I., Franzosi, M., Bonomi, G., *Mater. Plast. Elastomeri* (1972) **38**(2), 142.
32. Rohrschneider, L., Fresenius, Z., *Anal. Chem.* (1971) **255**(5), 345.
33. Levy, R. L., Wolf, C. J., Oro', J., *J. Chromatogr. Sci.* (1970) **8**, 524.

34. Ligon, W. V., Jr., Johnson, R. L., Jr., *Anal. Chem.* (1976) **48**(3), 481.
35. Grayson, M. A., Wolf, C. J., *Ann. Conf. on Mass Spectrom. Allied Topics, 25th,* Washington, D. C., 1977, Paper # **We43.**
36. Levy, R. L., Wolf, C. J., Grayson, M. A., Gibert, J., Gelpi, E., Updegrove, W. S., Zlatkis, A., Oro', J., *Nature* (1970) **227**, 148.
37. Levy, R. L., Grayson, M. A., Wolf, C. J., *Geochim. Cosmochim. Acta* (1973) **37**, 467.
38. Kausch-Blecken Von Schmeling, H. H., "Reviews in Macromolecular Chemistry," G. B. Butler, K. F. O'Driscoll, M. Shen, Eds., p. 97, Marcel Dekker, Inc., New York, 1970.
39. Devries, K. L., Roylance, K. K., Williams, M. L., *J. Polym. Sci. Part A-1,* (1970) **8**, 237.
40. Reich, L., Stivala, S. S., "Elements of Polymer Degradation," p. 1, McGraw Hill, Inc., New York, 1971.
41. Zhurkov, S. N., Tomashevskii, E. E., "Conference Proceedings: Physical Basis of Yield and Fracture," p. 200, Oxford, 1966.
42. Zhurkov, S. N., Zakrevskii, Va. A., Korsukov, V. E., Kuksenko, V. S., *J. Polym. Sci. Part A-2* (1972) **10**, 1509.
43. Amelin, A. V., Pozdnyakov, O. F., Regel', V. R., Sanfirova, T. P., *Fiz. Tverd. Tela* (1970) **12**(9), 2528. [*Sov. Phys.-Solid State (Engl. Transl.)* (1971) **12**(9), 2034].
44. Kamerbeek, G. H., Kroes, H., Grolle, W., *Soc. Chem. Ind. Monogr.* (1961) **13**, 357.

RECEIVED November 10, 1977.

6

The Characterization of Organic Polymers Via Pyrolysis–Infrared

W. L. TRUETT

Wilks IR Center/Foxboro Analytical, South Norwalk, CT 06856

The characterization of organic polymers is investigated by means of the pyrolysis–IR technique, which has been widely used in the past to identify intractable organic polymers. A series of polybutadiene polymers containing varying amounts of cis, trans, and vinyl groups have been decomposed at temperatures ranging from 500°–1000°C. The principle off gases are ethylene, propylene, and 1,3-butadiene, and band ratios can be used to determine the microstructure of the original polymer. In a related study, the pyrolysis–IR technique was applied to a series of ethylene–propylene copolymers. The amount of ethylene produced by the pyrolysis is proportional to the concentration of ethylene in the ethylene–propylene copolymer.

Pyrolysis–IR is the name applied to the technique in which a substance is pyrolyzed and then the products are characterized by IR spectroscopy (*1*). The chief use of pyrolysis in the past has been to characterize intractable organic polymeric substances (*2*). The early work, dating from 1953, was followed by many other investigations, based principally upon characterization of the liquid products from the pyrolysis of polymers and the very meaningful gaseous products were frequently neither collected nor characterized. The considerable advances which have occurred in IR spectrophotometers in the quarter century that has elapsed since the initial investigation now permits the rapid identification and quantitation of complex gaseous mixtures; this is owing to the fact that the present-day IR monochromator is based on the grating rather than the prism, thus the resolution is in terms of a wavenumber or less over a range of 2–50 μ.

0-9412-0406-3/79/33-174-081$05.00/1

Pyrolysis was performed using the Wilks Model 40 Pyrolyzer. In this device the polymer is placed on a filament contained in a small IR gas cell fitted with sodium chloride windows. Initially the device is charged with 10 mg of polymer, evacuated, and the filament temperature quickly elevated to the desired level, normally in the range of 500°–1000°C for a period of time sufficient to completely volatilize the sample, generally 30–45 sec. This pyrolysis device, which has been described in detail (*4*), enables one to examine all of the volatile fractions from the pyrolysis reaction. This fact combined with the high-resolution capability of modern grating spectrometers enables one to identify the gaseous products formed from a study of the IR spectrum without conducting a separation process. All spectra were determined using either a Perkin–Elmer 710B or 567 spectrophotometer.

A related work using pyrolysis–gas chromatography (PGC) has indicated that it is not only possible to identify the gross structure of polymers, but it is also possible to determine the microstructure. An example is the determination of the microstructure of polybutadiene polymers containing varying amounts of vinyl linkages (*3*). The polymerization of butadiene can proceed in a 1,4-fashion, in which the residual unsaturation is either of the cis or trans type, or in the 1,2-fashion in which the residual unsaturation is of the vinyl type. It was shown that the fraction of vinyl linkages in polybutadiene could be determined by PGC by measuring the ratio of ethylene to 1,3-butadiene gases produced from the pyrolysis (*3*) (*see* Table II).

In this work the series of polybutadiene polymers listed below in Table I containing varying percentages of cis, trans, and vinyl groups have been investigated by pyrolysis–IR to determine if this method could generate results comparable with the above noted study by PGC. The structure of these polymers was determined using NMR techniques. These polymers were not studied by PGC, thus the comparison of this work with that of Perry's (*3*) is a general one.

Table I. Composition of Polybutadiene Samples

Sample No.	*Trans (%)*	*Vinyl (%)*	*Cis (%)*
A	0.7	1.0	98.3
B	4.5	3.8	91.7
C	50.4	9.3	40.3
D	39.2	34.2	26.2
E	35.8	42.1	22.1
F	33.0	46.7	20.3
G	1.0	85.4	13.5
H	24.6	62.3	13.1
I	5.5	94.5	0.0

Table II. Pyrolysis Gas Chromatography Product Analysis of Polybutadienes

Vinyl Group (%)	*Ethylene/Butadiene Ratio*
18	3.3
32	4.0
63	5.3
79	7.0

The pyrolysis of polybutadiene samples A–I listed in Table I was conducted over a range of temperatures from 450° to 1000°C. The study showed that definitive information with respect to the microstructure of polybutadiene can be obtained via pyrolysis–IR. The most obvious effect determined in the pyrolysis of the polybutadiene was the change in gaseous products formed with increasing temperature. At 450°C, all types of polybutadiene polymers yielded a product mix of approximately 10% methane, 10% ethylene, 80% 1,3-butadiene, and a trace of acetylene. Interpretations of the IR spectra were based on spectral collections (*5*) and the work of Ballamy (*6*). Experiments conducted at 600°, 800°, and 1000°C showed a progressive shift toward more methane, ethylene, and acetylene in the gaseous mixture while the formation of 1,3-butadiene was reduced greatly. Thus, at 1000°C, the approximate product mix was 20% methane, 20% ethylene, 20% acetylene, and 40% 1,3-butadiene. It also should be noted that there are unidentified gases present in the products at 1000°C.

These results should be compared with the earlier pyrolysis mass spectroscopic studies of polybutadiene by Madorsky (*7*) over the temperature range from 325° to 400°C. Gaseous products that constituted the major components were ethylene, ethane, propylene, 1,3-butadienes, and butenes. No acetylene or methane were detected in this study, but a similar shift in product mix was noted with increasing temperature; viz, ethylene, ethane, and propylene became the major constituents, and 1,3-butadiene decreased by 50%. The precise structure of the polybutadiene investigated by Madorsky was not described. It should, of course, be noted that minor changes in the manner in which the pyrolysis of a polymer is carried out can cause a major change in the gaseous products formed. An excellent example of this is found in the work of Bailey, concerning the pyrolysis of polypropylene (*8*). In addition to the five major products isolated by Madorsky (*7*) it also was noted that the pyrolysis of polybutadiene yielded significant quantities of propane, butanes, pentenes, and pentanes. Using standard spectra (*5*), a search was made for these components in all gas mixtures in this study. None of these species could be identified; however, the extinction coefficient of many of these gases is low and the bands show considerable overlap even with the grating spectrophotometer, thus small quantities could be missed.

The previous study by Perry (3) had shown that PGC of a series of polybutadienes containing varying percentages of vinyl linkages gave products which permitted structural characterization of the polymer (*see* Table II).

The present pyrolysis–IR study under our conditions produced similar results (*see* Table III). An examination of the product yields, as the fraction of vinyl linkages increased over the range from 10 to 94.5%, showed an increase of ethylene relative to 1,3-butadiene. All of the data for Table III were collected under identical experimental conditions—600°C and 45 sec firing time. Essentially all polymer was completely vaporized under these conditions leaving virtually no residue. The ratio values were calculated from absorbance values of the peaks assigned to ethylene and butadiene, respectively. In addition, it was noted that the relative amount of methane also increased with increasing vinyl content of the polybutadiene.

The samples in Table IV are arranged in order of decreasing cis content. A large group of spectra obtained at 600°C were searched in order to determine the presence of absorbance bands diagnostic of cis content. An absorption band at 1000 cm^{-1} correlates well with the fraction of cis structure; however, it may also correlate in part with the presence of trans unsaturation. It is possible that under the conditions of the pyrolysis that a cis–trans equilibrium is reached. No absorption band could be identified with trans unsaturation, but the absence of a 100% trans sample made the identification difficult. Table V lists those diagnostic absorption bands of greatest utility in the identification of the gaseous pyrolysis spectra.

Earlier work (9) indicated that the addition of 1 mg of zinc dust exerted a catalytic effect upon the decomposition of vinyl polymers. One sample in Table I, G, was investigated extensively over the temperature range from 300° to 600°C. Addition of zinc dust lowers the temperature necessary to achieve pyrolysis of the polybutadiene. Zinc-catalyzed results obtained at 300°C compare favorably with uncatalyzed results achieved at 450°C. Normally 300°C does not give sufficient pyrolysis of polybutadiene to permit identification of products by IR spectroscopy. It is possible, of course, that the effect of the zinc was merely better conduction of heat to the sample and that no genuine catalytic effect was observed. However, zinc has a long history as a catalyst for the degradation of natural products and a single experiment with aluminum showed no such effect.

An allied study aimed at determining the gross composition of an ethylene–propylene copolymer was carried out to determine the feasibility

Table III. Pyrolysis–IR Study Results

	Vinyl (%)	*Ethylene/Butadiene Ratio*
A	1.0	0.5
B	3.8	0.5
C	9.3	0.8
D	34.2	3.2
E	42.1	3.2
F	46.7	3.5
G	85.4	7.6
H	62.3	4.2
I	94.5	8.0

Table IV. Determination of the Presence of Absorbance Bands Diagnostic of Cis Content

Sample	*Cis (%)*	*Relative Absorbance Values at 1000 cm^{-1}*
A	98.3	4.0
B	91.7	4.0
C	40.3	2.5
D	26.2	2.0
E	22.1	2.2
F	20.3	2.0
G	13.5	1.5
H	13.1	1.7
I	0.0	0.5

Table V. Absorption Bands Used for Spectral Identification

Methane	1300 cm^{-1}
1,3-butadiene	910, 1020 cm^{-1}
Ethylene	950 cm^{-1}
Acetylene	730 cm^{-1}
Propylene	910, 990 cm^{1-}

Table VI. Results of Pyrolysis on Samples of Ethylene–Propylene Copolymers

Ethylene (%)	*Ethylene–Propylene Copolymers Relative Absorbance 950 cm^{-1} Band*
10	1.0
23	2.6
36	4.0

of using pyrolysis–IR for this type of analytical problem. The distribution of monomer sequences along the chain was not the objective of this work, nor was the sequence for this particular sample known. The compositions studied contained approximately 10, 22, and 36% ethylene, the remainder being propylene except for a small percentage of diene cross-linking agent. The diagnostic IR absorption bands used in this study, which was carried out at 500°–800°C using the same apparatus as used in Part 1 of this study, were those listed in Table V for ethylene and propylene since no other gases were present in detectable quantity.

Pyrolysis on the above three samples of ethylene–propylene copolymers was carried out over a range from 500°–800°C using 10 mg of polymers. It was found that 700°C and 30 sec firing time completely volatilized the samples and all data was collected under these conditions. Table VI summarizes the findings.

Literature Cited

1. Truett, W. L., *Am. Lab.* (1977) June.
2. Harms, D. L., *Anal. Chem.* (1953) **25**, 1140.
3. Perry, S. G., *J. Gas Chromatogr.* (1967) **5**, 77.
4. Cassels, J. W., *Appl. Spectrosc.* (1968) **22**, 477.
5. "Hazardous Gases and Vapors: Infrared Spectra and Physical Constants," Beckman Instruments, Inc., Fullerton, CA, 1974.
6. Bellamy, L. J., "The Infrared Spectra of Complex Molecules," Wiley, New York, 1964.
7. Madorsky, S., "Thermal Degradation of Organic Polymers," Wiley, New York, 1964.
8. Bailey, W. J., *Soc. Plast. Eng. Trans.* (1965) **5**, 1.
9. Truett, W. L., FACSS Conference, Philadelphia, PA, 1976, Paper No. 331.

RECEIVED November 10, 1977.

7

Inelastic Electron Tunneling Spectroscopy of Coupling Agents

H. T. CHU, N. K. EIB, A. N. GENT, and P. N. HENRIKSEN

Department of Physics and Institute of Polymer Science,
University of Akron, Akron, OH 44325

Inelastic electron tunneling spectroscopy (IETS) gives results analogous to IR and Raman spectroscopy for monolayers or submonolayers of an organic substance absorbed onto a solid surface. 4-Hydroxybenzoic and 4-aminobenzoic acid have been proposed as coupling agents to adhere polymers to inorganic materials. IETS spectra have now been obtained for both these acids chemisorbed onto alumina. They are compared with IR and Raman results for the same and related materials to establish the nature of chemisorption. The evidence suggests that bonding is ionic in character (COO^- Al^+) with the phenyl ring oriented perpendicular to the surface. IETS results for model silane coupling agents have so far shown only that chemisorption does not occur on alumina at room temperature.

Inelastic electron tunneling spectroscopy (IETS) is a promising new method for investigating the vibrational spectra of organic molecules absorbed onto solid surfaces (*1, 2*). It uses quantum-effect tunneling of electrons through a thin insulating layer sandwiched between two metal films. The insulating layer should be approximately 2–3 nm (20–30 Å) thick for optimum resolution, but it can be composed of two parts, typically the oxide of one metal substrate and a superimposed monolayer or sub-monolayer of the organic substance of interest. Small voltages are applied across this insulating layer, and the vibrational spectra of its components are determined by the energy lost by tunneling electrons. Data is generally obtained in terms of the second derivative (d^2I/dV^2) of the tunneling current I as a function of the applied voltage V because

0-8412-0406-3/79/33-174-087$05.00/1

this is the IETS analog of an IR spectrum. IETS peak positions along the voltage axis correspond to vibrational mode energies (1 ev = 8065 cm^{-1}). However, areas under IETS peaks are affected not only by the amounts and intrinsic absorptivities of chemical bonds in the insulating layer but also by the orientation of these bonds relative to the applied electric field (*3, 4*). Both IR and Raman-active modes are observed in IETS with roughly equal intensity (*5*).

An outstanding feature of this new spectroscopic technique is its high sensitivity—materials present only as a fraction of a monolayer (as low as 10%) can be studied. Thus, it is well suited for examining chemisorbed species at interfaces. Up to now it has been applied mainly to small molecules such as formic and acetic acid (*6, 7*), amino acids (*5, 8*), and phenol (*9*), although some larger molecules such as DNA and RNA have been investigated also (*5*). Recent review articles (*10, 11, 12*) contain further details.

One potentially fruitful field of investigation is the use of chemical coupling agents to link polymer coatings to inorganic substrates. Both the type and amount of interfacial bonding are in principle accessible to direct observation by IETS techniques. In this exploratory study, we have examined chemisorption of two simple coupling agents, 4-hydroxybenzoic acid and 4-aminobenzoic acid (*13*), on alumina.

Experimental

Glass microscope cover slides are cleaned and dried, and four indium tabs are soldered onto one face (Figure 1); these are for attaching wire leads later. The cover slide is then placed in a vacuum evaporator, and pure (99.999%) aluminum is thermally deposited on it 100–400 nm thick

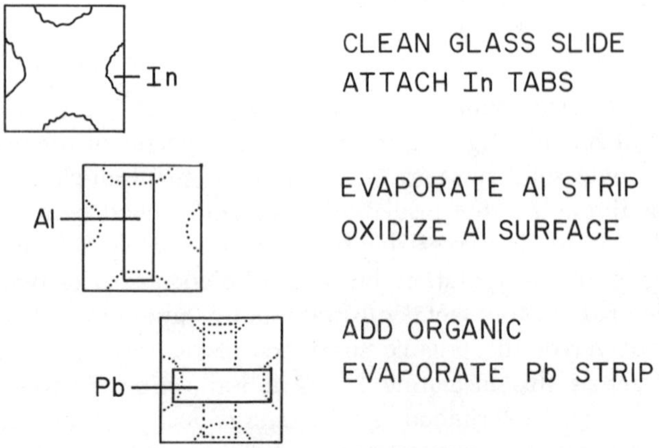

Figure 1. Sample preparation

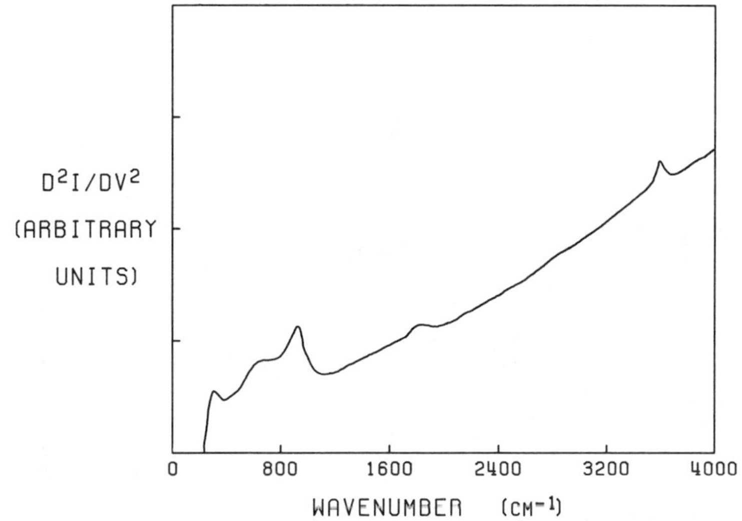

Figure 2. IETS results for a control sample (no organic material present on the alumina layer). Note the rising background from elastic tunneling processes. (See Table I for peak assignments).

through a mask to create the first electrode. This aluminum layer is then oxidized by a glow discharge method (*14*). Oxygen is admitted to the evaporator chamber at 0.05 torr, and the aluminum layer is exposed to a glow discharge produced by an electrode maintained at − 600 V dc with respect to the aluminum for 5–10 min. This technique produces a uniform oxide layer of controllable thickness.

The organic material of interest is applied next. A drop of a dilute (e.g., 0.005*M*) solution is placed on the oxidized aluminum surface, and the excess is immediately removed by spinning the sample rapidly. Water is a suitable solvent because it does not add additional peaks to the observed spectra. The sample is then evacuated again, and the upper metal film (lead) is deposited thermally through another mask at right angles to the aluminum strip to complete the junction (Figure 1). Wire leads are then attached, and the sample is inserted into a liquid helium Dewar at 4°K.

The circuit used to determine d^2I/dV^2 as a function of the applied voltage is a modified Wheatstone resistance bridge (*15*). The sample is used as one arm of the bridge, and a lock-in amplifier serves as the sample current detector. A small (2 mV) ac voltage is applied across the insulating layer at a frequency of 500.00 Hz, and the detector amplifier is tuned to twice this frequency, i.e., 1000.00 Hz, to determine d^2I/dV^2 (*15*). A dc voltage is applied in addition to the ac voltage to study the variation of d^2I/dV^2 with applied voltage. This dc voltage is cycled slowly over the range 30–500 mV corresponding to a wavenumber range of 240–4000 cm^{-1}. Multiple passes are made and the data subjected to signal-averaging to minimize random fluctuations. Some representative spectra are shown in Figures 2, 3, and 4.

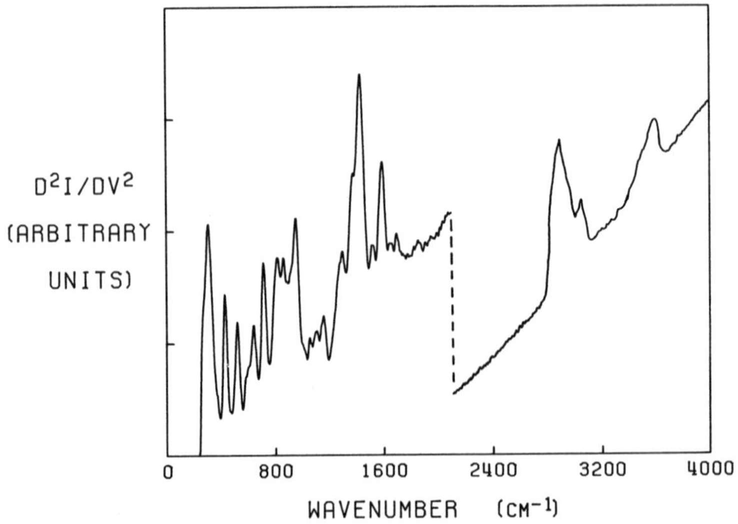

Figure 3. IETS results for 4-hydroxybenzoic acid chemisorbed on alumina. (See Tables II and IV for peak assignments). A vertical shift of the experimental relation has been imposed at 2100 cm^{-1}.

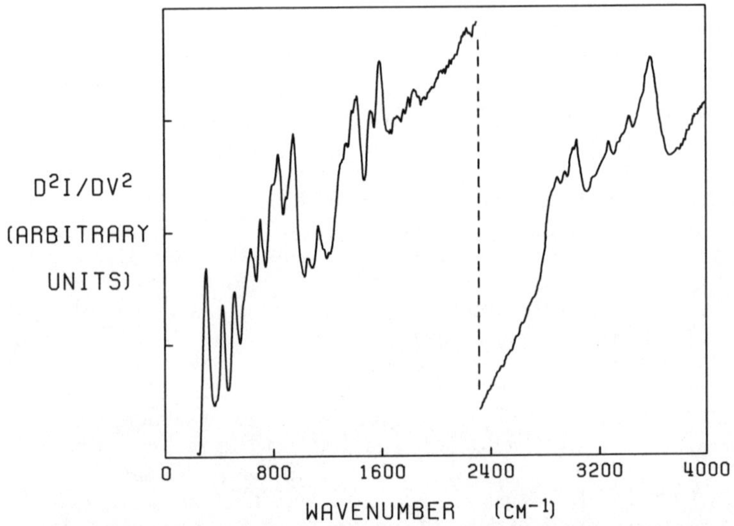

Figure 4. IETS results for 4-aminobenzoic acid chemisorbed on alumina. (See Tables III and IV for peak assignments). A vertical shift of the experimental relation has been imposed at 2300 cm^{-1}.

Results and Discussion

The IETS spectrum for a control sample having no organic material on the alumina layer is shown in Figure 2. Five peaks are observed superimposed on a smoothly rising curve resulting from elastic tunneling processes. Assignments for these peaks are listed in Table I. The peaks are relatively small and broad compared with peaks arising from chemisorbed substances (Figures 3 and 4), and they therefore generally do not interfere with other peak assignments.

IETS spectra for two coupling agents, 4-hydroxybenzoic acid and 4-aminobenzoic acid, chemisorbed on alumina surfaces are shown in Figures 3 and 4. Peak positions and indications of the relative peak heights are given in Tables II, III, and IV after subtracting a small correction term (7 cm^{-1}) from the observed peak positions to allow for the energy gap of the superconducting lead electrode (*16*). In Tables II and III these spectra are compared with published IR and Raman spectra for closely related compounds (*17, 18, 19, 20, 21*) to assign IETS peaks to specific bond motions as completely as possible. However, several weak combination peaks in the range 1700–2000 cm^{-1} have been omitted from the IR data for clarity. No equivalent peaks were detected by IETS.

Before comparing these spectra further, certain features of the published IR and Raman spectra should be noted. Ref. *17* deals with the alkali metal salts (Li, K, Na, Rb, Cs) of benzoic acid and salicyclic acid (2-hydroxybenzoic acid). Peak positions for the various salts of both acids were closely similar. Ref. *18* deals with four rare earth salts of benzoic acid (Pr, Nd, Er, Tm). Again, the peak positions were closely similar. Peaks have therefore been assigned by the authors to specific bond motions common to all of these substances; their assignments are given in the last columns of Tables II and III.

IR and Raman spectra for benzaldehyde chemisorbed on alumina were obtained using finely divided γ-alumina (280 m^2/g) treated with benzaldehyde vapor and heated to 280°C to remove any excess (*19*). IR

Table I. IETS Peak Positions for a Control Sample (with No Organic Material)

Peak Position (cm^{-1})[a]	*Assignment* (3–9)
319m	Al phonon
620b	OH bend
940m	Al oxide phonon
1870w	2 × Al oxide phonon
3620m	OH stretch

[a] Symbols: m, medium; w, weak; b, broad.

Table II. IR, Raman, and IETS Peak Positions (cm^{-1}) for 4-Hydroxybenzoic Acid and Related Substances[a]

Alkali Metal Salts of Benzoic Acid (17) (mean values)		*Aluminum Benzoate (18, 19) and Benzaldehyde on Alumina (19) (mean values)*		*4-Hydroxybenzoic Acid on Alumina*	*Assignment*
IR	*Raman*	*IR*	*Raman*	*IETS*	
				3660b	ν(Al–OH) (*3–9*)
3087w	3074w	3070w	3075m		
3070w		3033w		3057w	ν(CH) (*17–20*)
3038w		2850w[b]	2960w[b]		
				1868b	2 × β(Al–OH) (*3–9*)
				1705m	—
				1661w	ν(C═O) (*18, 20, 21*)
1604s	1600s	1600s	1605s	1600s	ν(CC) (*17–21*)
1595s					
1555s	1544m	1557s		1530m	asymν(COO^-) (*17–19, 21*)
1495w		1495m			ν(CC) (*17–21*)
1415s		1450s	1450m		
1390s	1392s	1430s	1430mw	1436s	symν(COO^-) (*17–19, 21*)
		1370m[b]		1384sh	—
1308w	1300w		1310w	1302m	ν(CC) (*17–20*)
1276w		1205m[b]	1210m	1284sh	
1177w		1172w	1165m	1161m	
1156w	1155m	1100b[b]		1110m	β(CH) (*17–21*)
1066m		1067m		1060m	
1020m	1023m	1020w	1030w	1032sh	
1002w	1001s		1005vs	995s	β(CCC) (*17–21*)
975w					
929w				922sh	
845m	843m		859m	864m	γ(CH) (*17–21*)
822m			820sh	820m	
707s				714s	γ(CCC) (*17, 18, 20, 21*)
682m					γ(CC) (*17*)
617w	617w		619m	638s	β(CCC) (*17–21*)
				596sh	
				581sh	β(Al–OH) (*3–9*)
535m				516s	γrock(COO^-) (*17, 18*)
				424s	βrock(COO^-) (*17, 18, 21*)
				302s	Al phonon (*3–9*)

[a] Symbols: vs, very strong; s, strong; m, medium; w, weak; sh, shoulder; b, broad; ν, stretch; β, in-plane bend; γ, out-of-plane bend; δ, scissors motion; (CH), aromatic C—H vibrations; (CC), aromatic C—C stretch; (CCC), aromatic C—C—C bend; (C═O), carbonyl stretch or bend; (CN), aromatic C—N vibration; (Al—OH), hydrated alumina stretch or bend; sym, symmetric stretch; asym, antisymmetric stretch.

[b] Only observed for benzaldehyde on alumina.

Table III. IR and IETS Peak Positions (cm^{-1}) for 4-Aminobenzoic Acid and Its Sodium Salt[a]

4-Aminobenzoic Acid (20, 21) IR (*mean values*)	*Sodium Salt of 4-Aminobenzoic Acid* (21) *IR*	*4-Aminobenzoic Acid on Alumina IETS*	*Assignment*
		3610s	ν(Al–OH) (*3–9*)
3476m	3359m	3450w	asym$\nu(NH_2)$ (*20, 21*)
3378m	3294m	3300w	sym$\nu(NH_2)$ (*20, 21*)
		3050m	
2980m		2975w	ν(CH) (*17–20*)
		2923w	
		1857b	2 × β (Al–OH) (*3–9*)
		1728w	—
1673s			ν(C=O) (*18, 20, 21*)
1638vs	1612m		$\delta(NH_2)$ (*20, 21*)
1602vs	1599s	1598s	ν(CC) (*17–21*)
1577s	1564m		
	1541vs	1534m	asym$\nu(COO^-)$ (*17–19, 21*)
1524w	1520sh		ν(CC) (*17–19*)
1446s			
1426s			δ(OH) (*20*)
	1403vs	1426s	sym$\nu(COO^-)$ (*17–19*)
1348m		1352w	ν(CC) (*17–19*)
1327vs	1325w	1310sh	$\beta(NH_2)$ (*20, 21*)
1290vs	1263w	1223vw	ν(CN) (*20, 21*)
1175w	1188m	1190sh	
1131m	1121w	1143m	β(CH) (*17–21*)
1071w	1084w	1066m	
1008w	1017w	995s	β(CCC) (*17–21*)
909m	884m	903w	$\gamma(NH_2)$ (*20, 21*)
841s	846s	844s	γ(CH) (*17–21*)
772s	781vs		β(CCC) (*21*)
699s	690m	708s	γ(CCC) (*17, 18, 20, 21*)
631m	638w	634s	β(CCC) (*17–21*)
619m	621w		
		580sh	β(Al–OH) (*3–9*)
552m	535m		$\gamma(NH_2)$ (*20*), β (C=O) (*21*)
		511s	γrock(COO^-) (*17, 18*)
499m	490m		γ(CCC) (*20, 21*)
	430w	423s	βrock(COO^-) (*17–21*)
390m	375m	381vw	wag(NH_2) (*20, 21*)
291w	300w	300s	γ(CH) (*20, 21*) + Al phonon (*3–9*)

[a] For symbol definitions, see Table II.

Table IV. IR and IETS Peak Assignments (cm^{-1}) for 4-Hydroxybenzoic Acid and 4-Aminobenzoic Acid[a]

4-Hydroxybenzoic Acid		*4-Aminobenzoic Acid*		
IR (KBr Pellet)	*IETS (On Alumina)*	*IR (KBr Pellet)*	*IETS (On Alumina)*	*Assignment*
	3660b		3610s	ν(Al–OH)
		3463m	3450w	asymν(NH_2)
		3367m	3300w	symν(NH_2)
	3057w	3048w	3050m	ν(CH)
		2977w	2975w	
			2923w	
	1868b		1857b	2 × β(Al–OH)
	1705m		1728w	—
1675vs	1661w	1671vs		ν(C=O)
		1622m		δ(NH_2)
1604s	1600s	1601s	1598s	ν(CC)
1594s		1571m		
	1530m		1534m	asymν(COO^-)
1505m		1515m		ν(CC)
1446m		1439s		
	1436s		1426s	symν(COO^-)
1420s		1418m		δ(OH)
1380sh	1384sh	1378w	1352w	ν(CC)
1314s	1302m			
		1311s	1310sh	β(NH_2)
1289s	1284sh	1287s		β(CH); ν(CN)
1244s		1238w	1223vs	
1167s	1161m	1171vs	1190sh	β(CH)
1123vw		1126w	1143m	
1099w	1110m	1115w		
	1060m	1081sh	1066m	
1010w	1032sh			
	995s	960sh	995s	β(CCC)
		879w	903w	γ(NH_2)
928m	922sh			γ(CH)
853s	864m			
834sh	820m	843m	844s	
769m		772s		β(CCC)
689w	714s	697m	708s	γ(CCC)
637w	638s	638w	634s	β(CCC)
606m	596sh	611m		
	581sh		580sh	β(Al–OH)
543w		546m		γrock(CO_2)
	516s		511s	γrock(COO^-)
504m		495m		γ(CCC)
	424s		423s	βrock(COO^-)
			381vw	wag(NH_2)
	302s		300s	γ(CH) + Al phonon

[a] For symbol definitions, see Table II.

results could not be obtained at wavenumbers below about 1000 cm^{-1} because alumina is insufficiently transparent under these conditions, and Raman results are unobtainable below 600 cm^{-1} because alumina fluoresces then. Within these limitations, the results agreed closely with the spectra of aluminum benzoate (*18, 19*) and other metal salts of benzoic acid (*17*) (Table II). From the almost complete disappearance of the characteristic aldehyde carbonyl (C=O) peak at 1690 cm^{-1} and from the appearance of symmetric and antisymmetric benzoate ion (COO^-) peaks at 1425 cm^{-1} and 1555 cm^{-1}, the authors concluded that benzaldehyde was chemisorbed on alumina as the benzoate ion (*19*). This conclusion was supported by the close agreement obtained between IR and Raman spectra for benzaldehyde chemisorbed on alumina and the corresponding spectra for aluminum benzoate (*18, 19*) and alkali metal benzoates (*17*).

The IETS results for 4-hydroxybenzoic acid chemisorbed on alumina are seen in Table II to be in generally good agreement with IR and Raman spectra for related compounds so that peak assignments can be made with some confidence. Moreover, they could be obtained down to 300 cm^{-1}, i.e., until phonon absorption in aluminum became dominant, in contrast to the more limited range of IR and Raman spectroscopy in this case.

In Table III, IETS results for 4-aminobenzoic acid chemisorbed on alumina are compared with published IR results for 4-aminobenzoic acid (*20, 21*) and for its sodium salt (*21*). Again, the IETS peak positions are in generally good agreement with the IR results.

In view of possible differences between the materials used for the present IETS studies and those used by other workers in their IR and Raman studies, IR spectra were obtained for the same samples of 4-hydroxybenzoic acid and 4-aminobenzoic acid used in the IETS experiments so that direct comparisons could be made. Samples in the form of potassium bromide pellets were examined over the range 500–4000 cm^{-1}; the results are given in Table IV together with the corresponding IETS results.

Slight but significant differences were found between the IR spectra obtained for the two acids and published IR spectra for related materials. These differences were generally not large enough to prevent peak assignments being made on the basis of the results given in Tables II and III. IETS results were in significantly better agreement with the newly determined IR spectra (Table IV) than with previously reported spectra (Tables II and III). The principal differences between the IR and IETS results are discussed below.

Only a small peak at 1660 cm^{-1} was observed by IETS for 4-hydroxybenzoic acid, and none was observed for 4-aminobenzoic acid where the peak for C═O would be expected, i.e., at 1670–1690 cm^{-1}. New peaks attributable to symmetric and antisymmetric stretching modes of the benzoate ion COO^- were found in both cases. Thus it may be concluded that these two acids are chemisorbed on alumina via the benzoate ion as was inferred previously for benzaldehyde (*19*).

The carbonyl C═O stretch mode gives rise to one of the most intense absorption peaks in IR. From the virtual absence of spectral peaks in the C═O position, it may be concluded that the amount of physical adsorption is small in the present cases. A small shoulder at 1685 cm^{-1} was observed for benzaldehyde adsorbed on alumina and was attributed to the carbonyl stretching mode in that instance (*19*). Some physical adsorption might occur by hydrogen bonding to the hydrated alumina surface (Al–O–H . . . O═COH–) or to benzoate ions on the surface (–COO$^-$ . . . HO–CO–).

Alumina is generally hydrated by a layer of hydroxyl groups separated far enough from each other to act independently (*22*). The OH stretch mode peak for the control sample (Figure 2) at 3620 cm^{-1} is fairly sharp, indicating that little hydrogen bonding occurs between

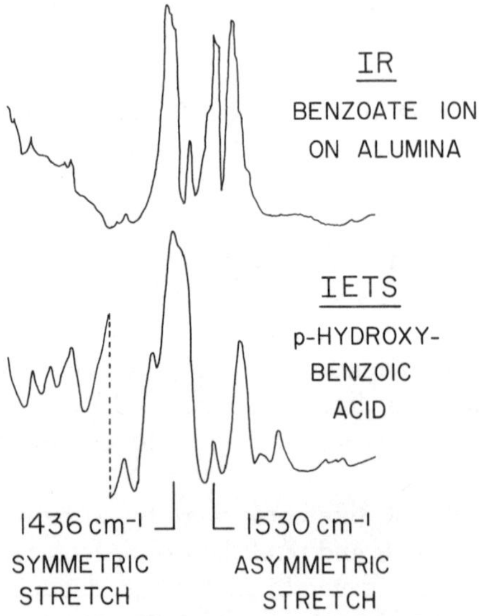

Figure 5. IR results for benzaldehyde on alumina (present as the benzoate ion) compared with IETS results for 4-hydroxybenzoic acid on alumina. A vertical shift has been imposed at 1300 cm^{-1}. Note the difference in intensity of the IR and IETS symmetric and antisymmetric stretch modes.

surface hydroxyl groups. However, these OH peaks in the presence of 4-hydroxybenzoic acid at 3660 cm^{-1} (Figure 3) and of 4-aminobenzoic acid at 3610 cm^{-1} (Figure 4) are significantly broadened, suggesting that the surface hydroxyl groups undergo hydrogen bonding, possibly to the acid ion ($-COO^{-} \ldots HO-Al$).

By comparing the IETS spectra of chemisorbed 4-hydroxybenzoic acid and 4-aminobenzoic acid with the IR spectra of chemisorbed benzaldehyde on alumina, conclusions may be drawn as to the orientation of the chemisorbed molecules. Symmetric and antisymmetric stretching modes of the benzoate ion are present with roughly equal intensity in IR as shown in Figure 5. In IETS, on the other hand, the symmetric stretching mode is much more prominent than the antisymmetric mode. This suggests that the molecule is oriented with the oxygens of the benzoate ion equidistant from the surface, as would be the case if the molecule is perpendicular to the alumina surface. With this configuration the antisymmetric stretching mode will largely involve motions parallel to the surface, i.e., perpendicular to the applied electric field, and thus it should not cause as strong an absorption as the symmetric stretching mode parallel to the field (*23*). Similar conclusions are reached for 4-aminobenzoic acid chemisorbed on alumina also. Thus, the two-coupling agents examined here, based on benzoic acid, appear to be chemisorbed on alumina in the form of benzoate ions with the oxygens lying equidistant from the surface, probably because the molecules are oriented perpendicularly.

Further experiments have been carried out with model silane coupling agents (trimethylmethoxysilane and vinyldimethylmethoxysilane) but so far with negative results. From the complete absence in the IETS spectra of the characteristic peaks (*24*), ν(Si–C) at 847 cm^{-1} and sym($Si-CH_3$) at around 1250 cm^{-1}, it was concluded that no significant degree of chemisorption of these silanes occurred under the experimental conditions used; i.e., when applied to alumina at room temperature as hydrocarbon solutions or as pure liquids. In view of the demonstrated sensitivity and wide spectral range of IETS, its use in examining coupling agents and other surface reactions of polymers seems well worthwhile.

Acknowledgments

This work forms part of a research program on the adhesion of polymers supported by a research grant from the Engineering Division of the National Science Foundation. The authors are also indebted to Richard Burgan of these laboratories for the design and construction of electronic apparatus and to P. Dreyfuss and M. L. Runge for helpful comments on interpretation of IR data.

Literature Cited

1. Lambe, J., Jaklevic, R. C., *Phys. Rev.* (1968) **165**(3), 821.
2. Geiger, A. L., Chandrasekhar, B. S., Adler, J. G., *Phys. Rev.* (1969) **188**(3), 1130.
3. Klein, J., Léger, A., Belin, M., Défourneau, D., Sangster, M. J. L., *Phys. Rev. B* (1973) **7**(6), 2336.
4. Simonsen, M. G., Coleman, R. V., *Phys. Rev. B* (1973) **8**(12), 5875.
5. Simonsen, M. G., Coleman, R. V., Hansma, P. K., *J. Chem. Phys.* (1974) **61**(9), 3789.
6. Skarlatos, Y., Barker, R. C., Haller, G. L., Yelon, A., *Surf. Sci.* (1974) **43**, 353.
7. Lewis, B. F., Mosesman, M., Weinberg, W. H., *Surf. Sci.* (1974) **41**, 142.
8. Magno, R., Adler, J. G., *Phys. Rev. B* (1976) **13**(6), 2262.
9. Lewis, B. F., Bowser, W. M., Horn, J. L., Luu, T., Weinberg, W. H., *J. Vac. Sci. Technol.* (1974) **11**(1), 262.
10. Brown, N. M., Walmsley, D. G., *Chem. Br.* (1976) **12**, 92.
11. Keil, R. G., Graham, T. P., Roenker, K. P., *Appl. Spectrosc.* (1976) **30**, 1.
12. Hansma, P. K., *Phys. Lett. C* (1977) **30**, 145.
13. Newbould, J., *J. Appl. Polym. Sci.* (1975) **19**, 907.
14. Miles, J. L., Smith, P. H., *J. Electrochem. Soc.* (1963) **110**(12), 1240.
15. Hebard, A. F., Shumate, P. W., *Rev. Sci. Instrum.* (1974) **45**(4), 529.
16. Kirtley, J., Hansma, P. K., *Phys. Rev. B* (1976) **13**(7), 2910.
17. Green, J. H. S., Kynaston, W., Lindsey, A. S., *Spectrochim. Acta* (1961) **17**, 486.
18. Taylor, M. D., Carter, C. P., Wynter, C. I., *J. Inorg. Nucl. Chem.* (1968) **30**, 1503.
19. Kuiper, A. E. T., Medema, J., Van Bokhoven, J. J. G. M., *J. Catal.* (1973) **29**, 40.
20. Théorêt, A., *Spectrochim. Acta* (1971) **27A**, 11.
21. Inomata, T., Moriwaki, T., *Nippon Kagaku Zasshi* (1970) **91**, 819.
22. Peri, J. B., *J. Phys. Chem.* (1965) **69**(1), 220.
23. Kirtley, J., Scalapino, D. J., Hansma, P. K., *Phys. Rev. B.* (1976) **14**(8), 3177.
24. Smith, A. L., "Analysis of Silicones," Wiley–Interscience, New York, 1974.

RECEIVED November 10, 1977.

8

Fourier Transform Infrared Spectroscopic Studies of Transitions in Polyethylene

WILLIAM W. HART and JACK KOENIG

Department of Macromolecular Science, Case Western Reserve University, Cleveland, OH 44106

FT-IR spectroscopic studies are performed on high-density polyethylene from 78 K to room temperature. IR bands indicative of crystalline and amorphous species reveal changes in their thermal responses at temperatures corresponding to well known relaxations. The sensitivity to these transitions depends on the IR-absorbing species and on the sample preparation conditions. The samples are thin films which have been slowly crystallized or rapidly quenched from the melt. Crystalline bands overall increase in intensity with increasing temperature probably because of increases in crystallite perfection and intermolecular forces. Superimposed on the increase of intensity are changes at temperatures corresponding to amorphous transitions. Amorphous bands exhibit changes in their temperature dependence primarily between 190 and 240 K. A dual glass transition exists for semicrystalline polyethylene.

Although perhaps the most extensively studied polymer, polyethylene (PE) is still the subject of controversy with respect to the nature of its thermal relaxation processes, especially the glass transition (T_g). The transitional phenomena of this polymer are the subject of an immense quantity of literature and several excellent reviewes (*1, 2, 3, 4*).

The temperature regime of particular interest is 100–300 K. Within these temperature limits several transitions occur which are only partially understood. The transitions occur at 250 ± 20 K, at 195 ± 10 K, and at 150 ± 5 K (*4*). The methods which have been used to observe these relaxations cover the entire frequency scale from dilatometry to electron spin resonance (ESR) and nuclear magnetic resonance (NMR). The

0-8412-0406-3/79/33-174-099$10.00/1

PE's studied had different degrees of crystallinity, molecular weights, and morphology. Copolymers of ethylene with propylene or vinyl acetate and aliphatic nylons were also investigated in efforts to separate morphological effects from the basic processes of the isolated methylene sequences.

The lowest temperature assigned to the glass transition (*1*) has been termed the gamma process. Torsion pendulum measurements (*1, 5, 6, 7*) suggest that this relaxation has two or more components. Dilatometry (*8*) resolved two discrete transitions at 108 and 152 K which have been labeled the γ_{11} and γ_1, respectively (*5, 6, 7, 9*). Illers' data on samples of various crystallinities support the conclusion that γ_{11} and γ_1 arise from crystalline and amorphous phases, respectively (*5, 6, 7*). Several mechanisms for this transition have been discussed which involve local motions of three to five methylene units (*5, 6, 7, 9*). The interpretation of this relaxation as the glass transition is doubtful (*2*).

The other two temperature regions of interest for the glass relaxations have been assigned by Boyer (*2*) as the lower (T_g(L)) and upper (T_g(U)) glass transitions. PE is only one example of the occurrence of a double glass transition in semicrystalline polymer systems. This phenomenon has been reviewed for the two-phase polymers by Boyer (*4*). One interpretation of the dual phenomena suggests that the crystallites serve as crosslink centers. Consequently, there are two forms of amorphous material with different mobilities: unstrained amorphous segments such as cilia and tie molecules or loose loops whose motions are hindered by the crystallites. These amorphous structures are thought to give rise to T_g(L) and T_g(U), respectively. NMR results support the existence of two distinct types of amorphous material (*10*).

T_g(L) is usually not observed in a direct fashion. Extrapolations of data to pure amorphous PE for samples of various crystallinities and ethylene content in copolymers have shown the glass transition to be at ca. 195 K (*2*). This transition has also been observed directly with dilatometry for bulk PE's by Lee and Simha (*8*). T_g(U) is readily seen by the thermal drift experiments of Chang (*11*) and volume relaxation measurements by Davis et al. (*3*). Dynamic mechanical measurements such as torsion pendulum require special conditions to observe a glass transition for PE. The transition at 190–250 K has been called anormolous friction since it can be enhanced by rapid cooling prior to observation (*12*). A relaxation is also revealed in this region after amorphous domains have been swollen by CCl_4 (*5, 11*). Both rapid cooling and swelling of the amorphous phase lead to creation of excess free volume in the amorphous areas (*13*). Therefore, these dynamic mechanical techniques allow the observation of an amorphous T_g free from the neighboring intense gamma relaxation.

Techniques which are more specific to the various morphological states, especially the amorphous domain, are needed. NMR and ESR are two such molecular probes. By monitoring the mobilities of protons as a function of temperature, Bergmann has defined the onset of molecular motion in various polymers (*14*). The applicability of NMR as a measure of molecular motion in polymer solids has been the subject of several reviews (*15, 16, 17*). ESR monitors the rotational and translational properties of stable radicals, usually nitroxides, and relates their mobilities to polymeric transitions. As stated in several works (*18, 19*), the radical's sensitivity to freedom of motion of the polymer chain is influenced by its size, shape, and polarity. The above probes are both high frequency in nature, 10^5–10^9 Hz. Measurement at high frequency has decreased resolving power for the various transitions in contrast to low frequency or static experiments, such a dilatometry with an effective frequency of 10^{-3} Hz (*20*).

One molecular probe, which, in theory, has the specificity and low frequency transitional resolution desired is infrared (IR) spectroscopy. Although this technique has been used to observe polymer transitions (*21–42*), it has not been used as often or with the success of NMR and ESR. An IR absorption's intensity, band shape, and frequency may change with temperature. There are two extremes of interpretation of the spectroscopy results. Either thermal expansion affects the inherent nature of the dipole moment change, or the concentration of the absorbing species changes with temperature. These two effects can occur simultaneously. This often precludes a straightforward analysis.

Fukawa demonstrated that the glass transition could be determined by monitoring the intensities of IR absorptions of polyvinyl chloride (PVC) (*21*). The intensity changes were similar to the temperature dependence of the refractive index and, consequently, to the dielectric constant. Anton recognized the relationship between the dielectric constant and the dipole moment (*25*). IR bands of higly polar groups showed a strong temperature dependence. Ogura et al. used this fact to study polyacrylonitrile with its highly polar nitrite vibration (*33*).

The sensitivity of highly polar groups to transitional phenomena observed by IR has some theoretical basis. Ovander showed by two theoretical treatments that a relationship could be derived predicting a linear dependence of the integrated intensity on temperature (*43, 44*):

$$I = I_o + \alpha T.$$

This equation had previously been found to hold for data of low-molecular-weight compounds (*45, 46*). The intermolecular expansion reduces the induced dipole moment of the interaction and therefore the intensity

(*45, 46*). The effect would be most pronounced for bands which arise from vibrational groups which are inherently very polar.

The IR type of molecular dilatometry should observe the same transitions as bulk thermal expansion measurements. An additional benefit may be achieved with IR studies of multiphase systems such as copolymers and semicrystalline systems. If certain absorptions can be related to the various components or morphological structures present, then a probe of the thermal responses of these structures is available. Koenig and co-workers used this rationale to observe IR absorptions of polyethylene terephthalate (PET) and polystyrene (PS) at cryogenic temperatures (*28, 34*). They attempted to isolate the morphologies responsible for particular transitions of PET with the knowledge of the structural character of certain IR bands. No concentration changes were expected in the low-temperature regions of these studies, and localized dilatometry effects were ascribed to all changes. In more recent studies of the liquid–liquid transition of PS in the melt, this laboratory has shown that Ovander's relation can be exclusively applied only when the samples have identical thermal histories (*42*). Otherwise concentration and kinetic effects obscure this approach.

If a thermal transition does cause a change in the distribution of various conformations or structures in the system, then the IR spectrum will reflect these concentration changes. Anton measured the IR absorptions sensitive to helix regularity, usually in the 900–1100 m^{-1} region, for several polymers as a function of temperature (*25*). Intensity decreases of these bands were observed above T_g. This change was attributed to a randomization of the chain structure. Anton illustrated this method of determining T_g for PET, polyvinyl acetate, PS, and several nylons. Other phenomena were observed at $T > T_g$ such as crystallization of PET and a crystal–crystal transformation of nylon 66.

Some specific structures have been investigated on the basis of concentration differences. Examples are the ratios of bonded and free hydrogen bonds in nylons (*25, 29, 30, 41*) and the extent of dimerization in methacrylic acid polymers (*26, 35, 38*) and their subsequent influence on such transitions as the melting point. Even at cryogenic temperatures, emphasis has been placed on concentration effects in certain studies. Belopolskaya et al. investigated the IR spectrum of polymethyl methacrylate of various tacticities at 65–450 K (*23, 32*). In the region of the ester vibration, 1000–1350 cm^{-1}, as many as seven vibrational components were isolated by curve-resolving techniques. Integrations of these components' intensities allowed the assignments of transitions seen with birefringence at 50 and at 300–400 K (depending on tacticity) to side group and main chain alterations, respectively. However, the separation of the concentrational and dilatometric effects on intensity changes is not always this clear.

Frequencies of absorptions also change with temperature. The positions as well as the intensities of the amide and N–H vibrations of nylons were studied extensively by Nikitina et al. over the temperature interval of −160°C to the melting point (*30*). The shifts were interpreted as strengthening of the hydrogen-bonded character of the structure as temperature was decreased.

Shifting of an IR peak frequency with temperature has been studied most thoroughly for the ethylenic polymers. However, these studies have been interested primarily in crystal variations with temperature rather than with transitional phenomena. The three principal vibrations in the mid-IR (methylene stretching, bending, and rocking) are split into doublets. This phenomenon is known as Davydov or factor group splitting (*47*). It can be viewed as the interaction of the two chains in the orthorhombic unit cell. Because of the close proximity of packing, intermolecular forces sufficiently perturb the vibrations to allow the observation of the doublets predicted by lattice symmetry. Since thermal contraction will change the crystal dimensions, the absorption frequencies are affected as the interchain interactions increase.

Brockmeier has studied this temperature effect at 22–313 K for PE and copolymers of ethylene with propylene (*48*). The results obtained by monitoring the frequencies of the three doublets could not be attributed to a simple contraction of the unit cell. London dispersion forces and changes in the helix of the molecule were possible complicating factors. Shen et al. have observed the shifting of the rocking doublet, ca. 725 cm^{-1}, from 10 K to room temperature for solution-crystallized PE (*49*). The 731 and 720 cm^{-1} components are polarized along the *a* and *b* axes of the unit cell, respectively. Shen found no preferential correlation of either frequency's shift with x-ray measurements of expansion of a crystalline dimension. However, a linear relation was found for the frequency of the 731 cm^{-1} band with the specific volume of the unit cell. In both IR studies, frequency shifts were minimal below approximately 50 K which Brockmeier suggested arose from the lower limit of molecular motion in the crystal (*48*).

Frank and co-workers also examined the IR spectrum of this polymer as a function of temperature to understand more clearly the mechanism of thermal contraction (*50*). However, their investigations were concerned primarily with the intensity and position of the 73 cm^{-1} band as a function of temperature from 14 K to the melting point. They concluded that the data of the 73 cm^{-1} absorption correlated well with previous x-ray measurements for expansion along the *a* axis of the unit cell, and an inflection was observed at ca. 243 K. This temperature would correspond, according to Frank et al., to the temperature at which diffusion of defects into the crystal can occur.

Recently, Hendra et al. have reported the ability to quench linear PE to the glassy state (*51*). The extent of splitting of the 725 cm^{-1} doublet was one method used to ascertain the onset of local ordering as the sample warmed. This could be done because the splitting arises from crystalline packing of the chains. In this manner it was concluded that the glass transition occurred below 190 K. Boyer and Snyder, however, integrated Hendra's spectra by separating the crystalline doublet from the amorphous background (*52*). The integrations yielded a drastic change from glassy or disordered to crystalline at 195 K. This affirmed Boyer's previous contentions regarding the glass transition temperature. Their results showed order developing above 160 K, although not as rapidly as above $T_g(L)$. Jones et al. have obtained glassy PE by quenching thin films supported on electron microscope grids (*53*). Their electron diffraction results also showed crystallization occurring below 190 K. DSC scans of the amorphous samples revealed an irreversible exotherm at about 160 K which was attributed to the onset of crystallization.

Crystalliaztion may occur by defect motion associated with the gamma transition. This motion, although small, could be enhanced by the high free volume of the quenched state. Long chain motion, normally associated with T_g, may not be necessary for the flexible ethylene chain to achieve some degree of local order. Additional information regarding the character of these transitions may be obtained from a study of the reversible temperature effects on the IR spectrum of PE.

The nature of the structures responsible for the various absorptions has been the subject of a plethora of spectroscopic research, both experimental and theoretical. Barnes and Fanconi have reviewed and recalculated the assignments of vibrations of PE and the *n*-alkanes (*54*). Inter- and intramolecular interactions in the perfect lattice are also included in the computations. The assigned bands relate to a perfect crystal of the all trans methylene conformation.

The morphology of the semicrystalline sample is not a perfect structure. Bands which are characteristic of amorphous conformations are also present. The most intense of the amorphous absorptions are the methylene wagging modes at 1369, 1353, and 1303 cm^{-1}. Snyder has compared these vibrations with those observed for liquid alkanes (*55*). After showing that the wagging mode was stronger for methylenes adjoining gauche (G) rather than trans (T) bonds, he calculated the frequencies corresponding to several types of gauche sequences and found that the GTG conformation correlated with bands observed at 1369 and 1303 cm^{-1}. The 1353 cm^{-1} absorption was assigned to the wagging of a methylene isolated by two gauche bonds.

Zerbi and co-workers calculated the contribution of defects to the IR spectrum when incorporated in a long trans sequence of 200 methylenes (*56*). Bands were calculated at 1352 and 1350 cm^{-1} for the GG and

GTG structures, respectively. The GTTG defect caused two bands at 1370 and 1320 cm^{-1}. Only the band assignment for the GG conformation is the same for both Snyder and Zerbi. The disagreement of the other calculated frequencies was attributed by Zerbi (*56*) to end effects of the model compounds used in Snyder's work.

The limiting frequency for the isolated methylene chain having the all trans conformation is 719–720 cm^{-1}. In actuality, this frequency is approached after six consecutive trans units (*57*). The 720 cm^{-1} peak arises from both amorphous and crystalline phases. Since the 731 cm^{-1} band originates from close-packed interactions, it is caused solely by the crystalline phase. This correlation of the rocking doublet and the wagging modes to the various morphological structures is quite useful. Monitoring of their respective intensities as a function of temperature should reveal an understanding of the molecular processes associated with thermal relaxations. No previous study has been performed using this approach.

Previous mid-IR spectroscopic studies of polymer transitions, with the exception of Enns et al. (*42*), have used conventional dispersive instruments. One narrow range of the IR spectrum was observed by slowly scanning the frequency as the temperature was varied. Until recently this has been the optimal experimental instrumentation. However, with the advent of the fast Fourier transform algorithm, a new dimension has been added in the form of Fourier transform infrared (FTIR) spectroscopy. Application of this instrumentation to polymer systems has been the subject of recent reviews (*58, 59*). Two texts offer a more in-depth survey of this technique in relation to conventional dispersive spectroscopy (*60, 61*).

The major advantages of FTIR for a temperature-dependent study are speed, sensitivity, and data processing. The entire mid-IR range (4000–400 cm^{-1}) is obtained with the same ease and time as one narrow frequency range by conventional dispersive instruments. This advantage becomes apparent when one is limited by the thermal stability of the cryostat or heater containing the sample. Therefore, the spectroscopist is no longer deterred by time or thermal fluctuations from observing the entire mid-IR range as a function of temperature.

When the dedicated computer of the FTIR performs the transform of the interferogram, the resultant data is in digitized form. This allows the numerical comparison of specimens of differing morphology and temperature, by such methods as digital subtraction. Integration and other forms of data treatment are also more tractable with the aid of the on-line computer. The other advantages such as a high signal-to-noise ratio resulting from high energy throughput, multiplexing, and co-addition of scans have, once again, been discussed elsewhere (*58, 59, 60, 61*).

The use of FTIR to study a polymer's response to temperature seems, for the above reasons, spectroscopically logical. Cryostats are now available which control temperature to within ±0.1 K. When these experimental improvements are coupled with the spectroscopic knowledge of PE, a better understanding of the molecular nature of its relaxation processes may be obtained.

Experimental

The linear PE studied was Marlex 6015 obtained from Phillips Petroleum Co. The polymer powder was compression-molded between Teflon-coated aluminum foil at 170°C. Temperature was maintained for 15 min prior to compression for 5 min. The sample was then allowed to cool slowly to room temperature with no pressure while still in the platens of the press. Thin films were obtained which showed no macroscopic melt flow orientation and which were less than 0.002 in. thick. This sample preparation will be referred to as slow-crystallized.

Samples from these films were again placed between Teflon-coated aluminum foil and were suspended in a forced convection oven at 170°C for 30 min. The molten polymer was then mechanically propelled into a bath of isopentane slurry at −160°C as described previously (*53*). The isopentane was cooled with an outer jacket of liquid nitrogen. The quenched sample was removed from the isopentane and was allowed to anneal at room temperature for 20 hr prior to examination by IR spectroscopy at 78–290 K.

After observation, this film was placed in a vacuum oven at 100°C for 10 hr. This sample history will be referred to as annealed. An additional sample was quenched and subsequently was annealed by the same method, but using ice water as the quenching medium. Although this sample was not studied cryogenically with IR, it allowed a comparison with the isopentane quenching method.

The cryostat used for the low temperature measurements was an Oxford Instruments DN704 shown in Figure 1. Cooling is accomplished by circulating liquid nitrogen from the reservoir through the heat exchanger. The exchanger is essentially a copper jacket surrounding the sample space. The nitrogen flow rate was controlled by manually adjusting the exhaust vent. Heating was performed by an electric resistor mounted to the heat exchanger. Power to this element was provided with a proportional controller manufactured by Love Controls Corp., model 49. Nitrogen gas in the sample compartment was used as the exchange medium between the heat exchanger and the sample. The film was held loosely between two brass brackets mounted at the end of the sampling rod. The outer and inner vacuum windows were KBr and KRS-5, respectively. Temperature was monitored by two copper–constantan thermocouples. One was located in the heat exchanger and was referenced to the liquid nitrogen reservoir. It was used in conjunction with the proportional controller. Another thermocouple was placed in the exchange gas next to the sample to eliminate any uncertainty in the exchange process.

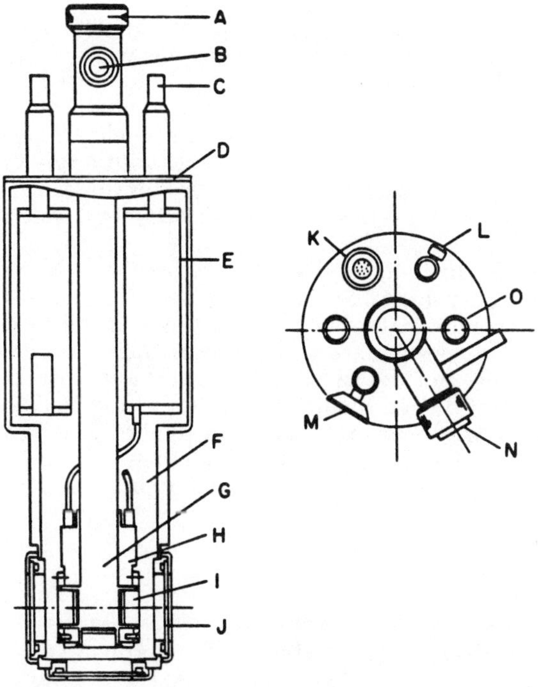

Figure 1. Liquid nitrogen cryostat manufactured by Oxford Instruments; model DN-704.

(A) O-ring sealing sample access port; (B) sample tube evacuation and safety valve; (C) nitrogen entry ports; (D) top plate also includes an inner evacuation valve, a 10-pin electrical connection, and a nitrogen exhaust valve; (E) nitrogen reservoir; (F) vacuum insulation space; (G) exchange gas and sample space; (H) heat exchanger and heater; (I) demountable inner vacuum windows; (J) demountable outer vacuum windows; (K) 10-pin electrical connection; (L) inner evacuation valve; (M) nitrogen exhaust valve; (N) sample tube evacuation and safety valve; (O) nitrogen entry ports.

Spectra were obtained using an FTIR instrument manufactured by Digilab, model FTS-14. The cryostat was plaed in the sample chamber and was enclosed by a Plexiglas shroud. This maintained the efficiency of the purge of dried air. The sample was mounted in the cold cell, and the temperature was lowered slowly to the base temperature of 78 K. Cooling time was normally 1 hr. The temperature was raised by increments of less than 10 K. An additional 10–15 min were allowed for the samples to equilibrate after the readings of the two thermocouples were similar. The spectrum was then taken at each temperature. The resolution used was 4 cm^{-1} in data acquisition. The frequency scale for FTIR measurements is internally calibrated with a reference He–Ne laser to an accuracy of 0.2 cm^{-1}.

The time to record spectra of acceptable quality in a temperature study is of basic importance. Several alterations were made in the standard method of obtaining spectra by double-beam FTIR instruments to reduce the measurement time. The sample was not moved, and the purge was maintained throughout the temperature cycle. Therefore, it was acceptable to sample in the single beam mode with a reference spectrum stored on computer. This halved the data acquisition time, thus increasing the efficiency and decreasing any uncertainty concerning thermal drift.

The energy to the detector was reduced greatly by the cryostat. The gain of the detector preamplifier was increased to enhance the measured signal. Signal averaging was, because of this alteration, more effective in increasing the signal-to-noise ratio. The amplification caused a reduction in the baseline but did not affect the absolute absorbances. These modifications allowed good-quality spectra to be obtained with 200 scans. This infers a measurement time of less than 10 min. The error of temperature measurement caused by thermal fluctuations was minimal, and temperatures are thought to have an accuracy of ± 1 K.

Results

The absorption bands of interest in the IR spectrum of PE are in the region 1425–600 cm^{-1}. Crystalline and amorphous phases contribute to vibrations observed within these limits (*62*). Amorphous methylene wagging vibrations are observed at 1369, 1353, and 1303 cm^{-1}. The broad peak at 1078 cm^{-1} has been assigned to the C–S stretching vibration of the amorphous phase.

The linear PE studied contains methyl end groups which are reflected in the methyl symmetric bending mode at 1376 cm^{-1}. This barely resolvable as a shoulder of the 1369 cm^{-1} amorphous band. The 909 and 990 cm^{-1} vibrations are characteristic of the terminal vinyl groups. These absorptions correspond to the out-of-plane and in-plane bending deformation of the C–H group. The associated dipole moment is stronger than those of the methylene deformations. Therefore, although very low in concentration, the vinyl absorptions appear clearly in the spectrum. This allows the comparison of samples of varying thickness and morphology. These peaks are illustrated in Figure 2 for a slow-crystallized sample measured at room temperature. Representative baselines are drawn for the intensities which were monitored as a function of temperature.

The 725 cm^{-1} rocking mode doublet is presented in Figure 3. The peaks at 731 and 720 cm^{-1} for this sample are approximately 15 times the intensity of the amorphous wagging modes. The 720 cm^{-1} limiting frequency is assigned to long trans sequences in the amorphous and crystalline phases. For these samples, most of the methylene structures are in the crystal, and the band is associated with this phase predomi-

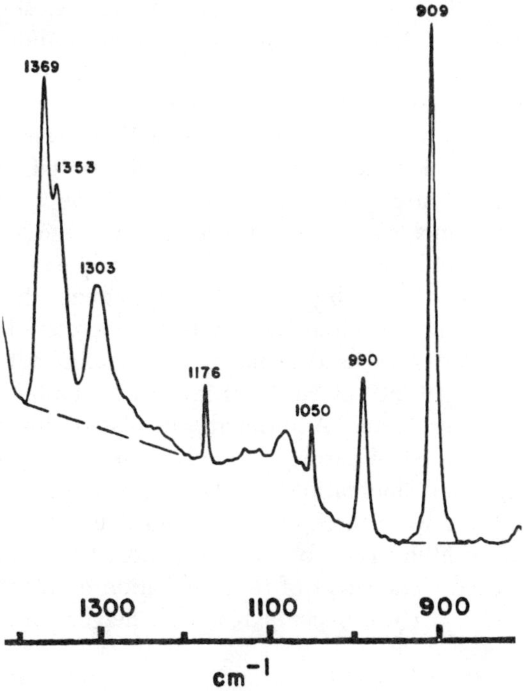

Figure 2. Slow-crystallized PE at room temperature

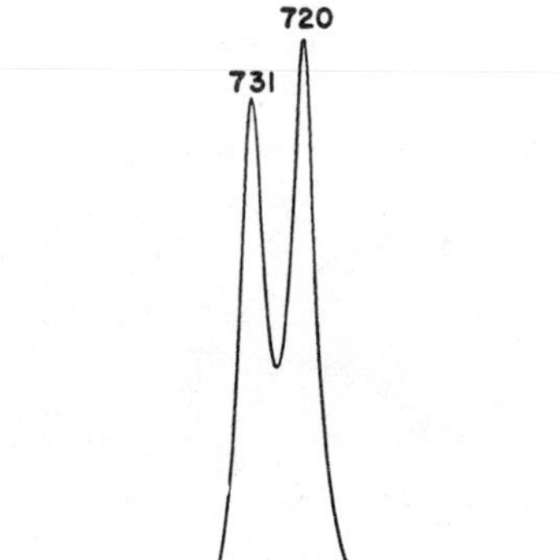

Figure 3. Slow-crystallized PE at room temperature

nantly. Since the 731 cm^{-1} vibration originates from close-packed intermolecular interactions, it is solely caused by the crystalline phase.

The spectra of a very thin film (ca. 10–15 μm) at the two temperature extremes of this study are shown in Figure 4. This sample was slow-crystallized from the melt. The dominance of the bending and rocking doublets over the weak amorphous motions is illustrated. At first observation, one notices a large increase in intensity of the band at 731 cm^{-1} as the temperature increases. The spectra for this region are expanded in Figure 5. The bands have shifted to higher frequency at lower temperatures. Since the data is digitized at predetermined frequencies, the shift results in the peak's intensity not being monitored at points of similar position relative to its true maximum as temperature is varied. A data handling capability of the instrument's computer software was used to solve the problem. Additional, calculated, interspersed data points were interpolated by fitting an eleventh degree polynomial to the real data. Only small fluctuations in the interpolated intensities were observed when there were three or more filled data points between real ones. Therefore, a filling ratio of three interpolated points to one real data point was used. The effect of the calculation on the spectra is also shown in Figure 5. Figure 6 compares peak height measurements for

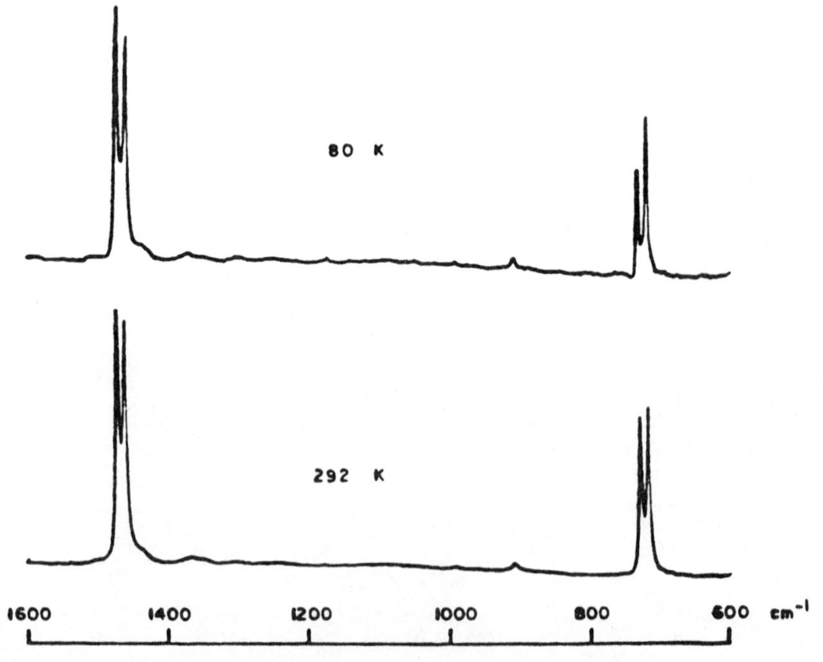

Figure 4. Comparison of spectra of slow-crystallized sample at temperature extremes

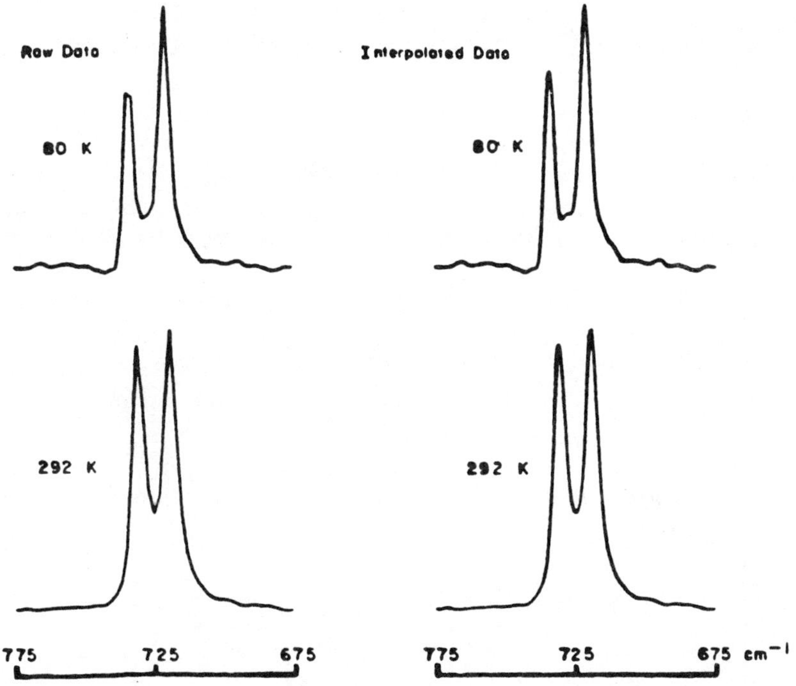

Figure 5. Comparison of raw and interpolated spectra for 725 cm⁻¹ doublet of slow-crystallized sample at temperature extremes

the raw and interpolated data over the temperature interval studied using this sample. The thermal transitions cited in the literature (*4*) also are noted. This result illustrates the effect of frequency shifts on the measurement of peak heights. Although the effect is most noticeable when the peak is narrow and the shift large, as with the rocking mode, smaller errors are incurred when observing the temperature dependence of othr bands. For this reason, all results will be interpolated spectra.

The various samples studied have differences in morphology which are reflected in the IR spectra. The region of 1425–800 cm^{-1} is presented in Figure 7. The spectra have been expanded in this plot so that the intensity of the vinyl end group vibration at 909 cm^{-1} is comparable for the spectra at 80 K. This has normalized the absorbance scale with respect to sample thickness. At the higher temperature, the same expansion was used.

The intensities of the vinyl absorption bands increase as the temperature is lowered. The crystalline bands at 1050 and 1176 cm^{-1} exhibit a much narrower bandwidth at lower temperatures. Only slight changes in the amorphous bands are observed with temperature. However, differences between slow-crystallized and quenched samples are apparent in this figure. The amount of amorphous material is much larger in the quenched sample. After annealing at 100°C for 10 hr, these structures decreased only slightly.

The same expansion factors were used to plot the data for the 725 cm^{-1} methylene rocking doublet (Figure 8). In every sample the 731 cm^{-1} absorption band decreases with temperature. At higher temperatures the relative intensities of the doublet for the quenched and the subsequently annealed samples are reversed compared with the slow-crystallized sample. The 731 and 720 cm^{-1} bands are highly polarized along the *a* and *b* crystal axes, respectively. It was suspected that nonrandom orientation with respect to the thickness direction was induced by the quenching process. Angood and Koenig have discussed this type of orientation on IR measurements (*63*). Only vibrations in the plane normal to the incident rediation are observed.

Deviations from random orientation are observed by changes in relative intensities if the sample is tilted. The tilting process was per-

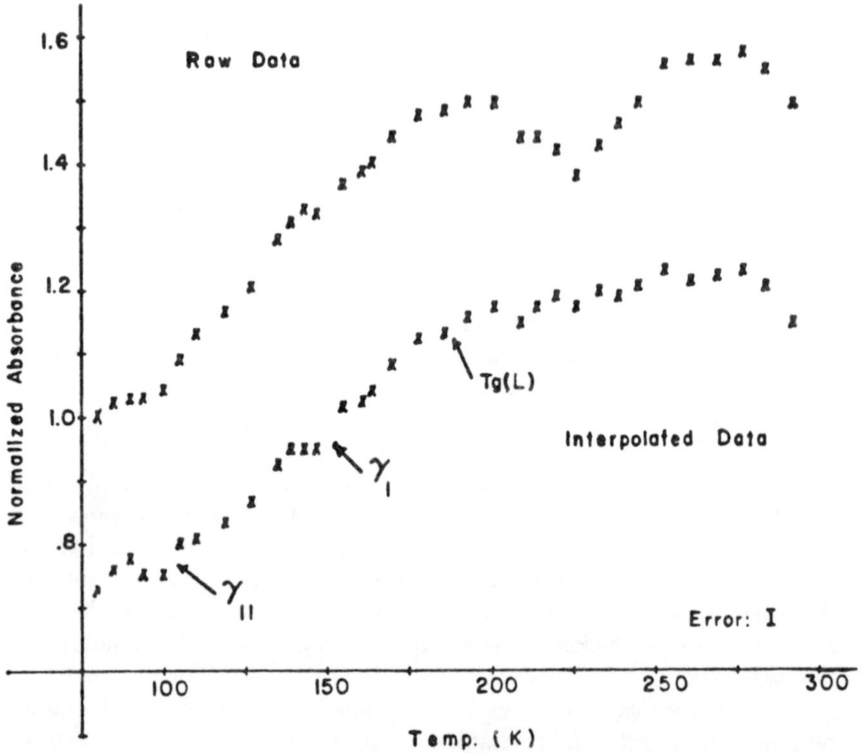

Figure 6. Normalized peak heights of 731 cm^{-1} absorption vs. temperature for raw data and interpolated data of slow-crystallized sample. Maximum error of measurement, 3%.

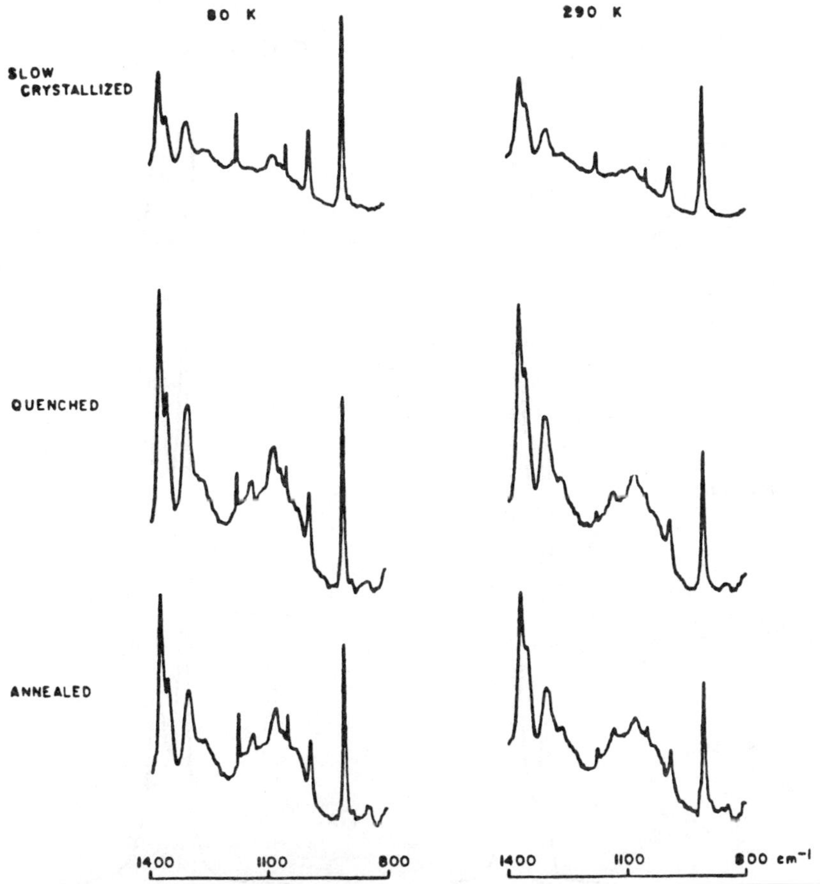

Figure 7. Effects of temperature and sample preparation on absorptions from 1400 to 800 cm⁻¹ in the spectrum of PE

formed for a quenched sample. The corresponding change in relative intensities of the 720 and 731 cm^{-1} bands is shown in Figure 9 and is, therefore, evidence for nonrandom orientation in the thickness direction. Similar orientation phenomena at surfaces has been observed by Schonhorn for crystallization at high energy surfaces (*64*). The growth direction of the crystal (*b* axis) was perpendicular to the surface of nucleation. This would tend to reduce the intensity of the 720 cm^{-1} band. Although the surface energy of the Teflon surface used in these studies was low, the effect can still be explained in terms of the rapid quench. Since a large thermal gradient was imposed normal to the surface, the growth of crystallites in this direction was enhanced.

Digital subtraction was used to illustrate further the effects of varying sample preparation and temperature. The 909 cm^{-1} vinyl band has

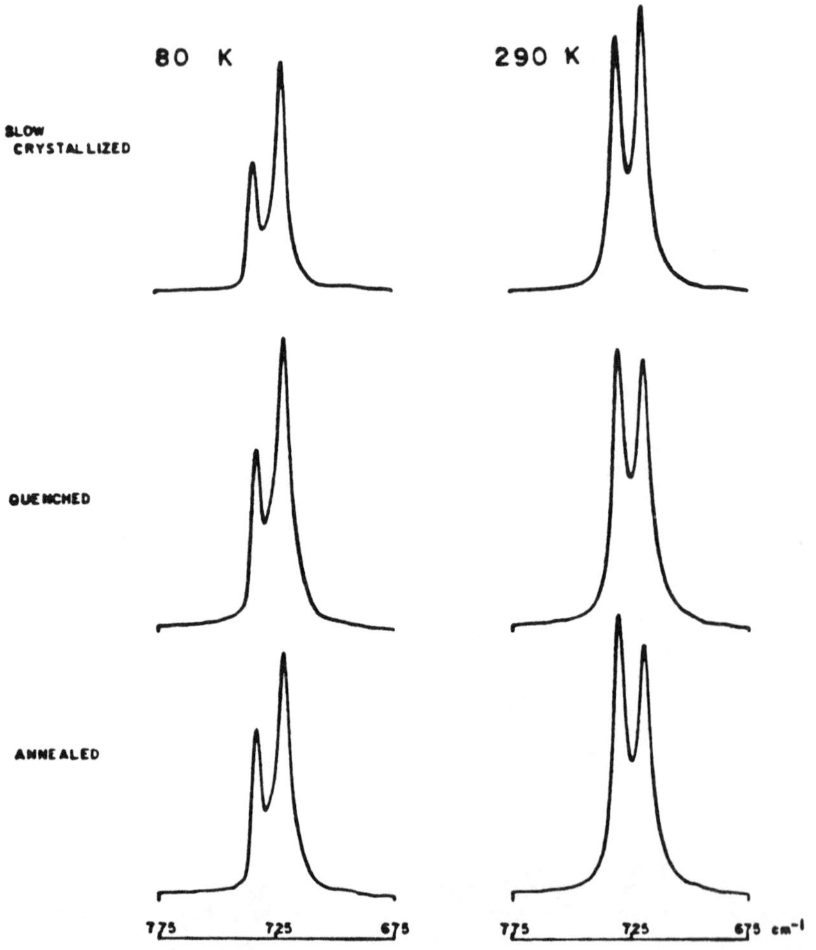

Figure 8. Effects of temperature and sample preparation on 725 cm^{-1} doublets of PE

already been used in Figure 7 to compare spectra of samples having different thicknesses. If it is used as the subtraction criterion or internal standard, it will allow a more rigorous comparison. After the internal standard vibration is subtracted from the baseline for the two samples, then the difference spectrum is considered independent of the relative sample thicknesses.

The method of digitally subtracting spectra is illustrated in Figure 10. This subtraction shows that the isopentane-quenched sample has twice as much amorphous material as the one which was slow-crystallized. The expansion factor for plotting of the difference spectrum rela-

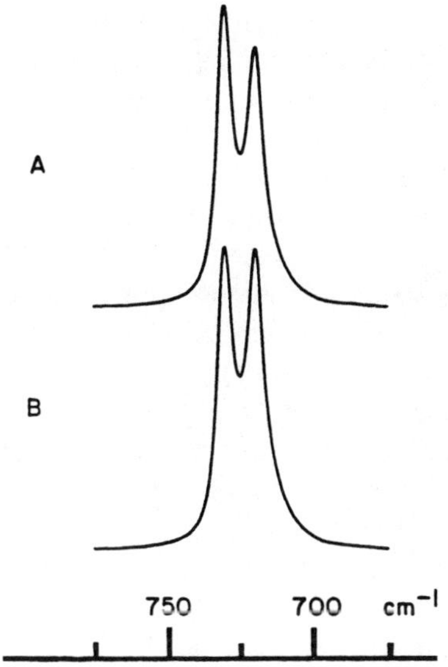

Figure 9. Room-temperature spectra of quenched sample. Position of film with respect to incident IR beam: (A), normal; (B), 45°.

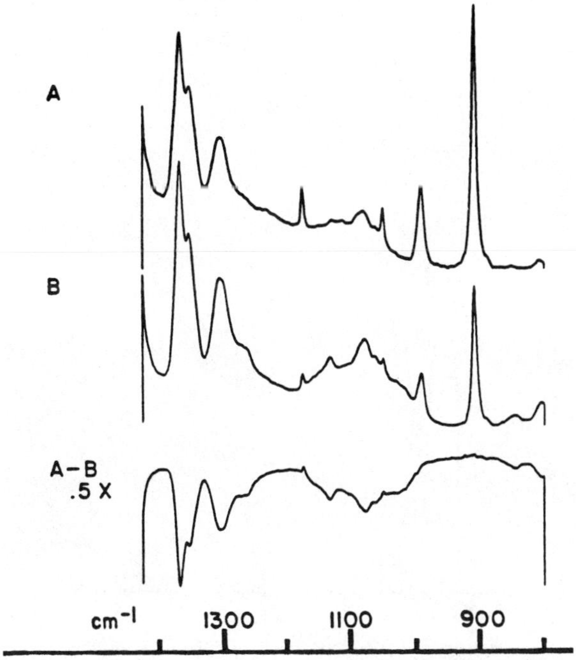

Figure 10. Subtraction of spectra at room temperature: (A), slow-crystallized; (B), quenched in isopentane slurry; (A − B), subtraction (1:1.2).

tive to the first spectrum is shown in the figure (0.5x). The subtraction coefficients to ratio the two spectra are contained in the caption (1:1.2). Figure 11 compares this difference spectrum with the one for the samples at 78 K. The same difference in morphology is seen here. The signal-to-noise ratio of the spectra decreased when taken in the cryostat. This is apparent in the difference spectrum and will be discussed later.

When the isopentane-quenched sample is annealed at 100°C for 10 hr, only a 10% decrease in amorphous structures is observed (Figure 12). The crystalline vibrations show only a slight increase. The difference spectra at 78 K show the same behavior (Figure 13). A sample was also quenched in ice water for comparison with the isopentane quench method. The subtractions of the spectra of the two quenched specimens from the low-crystallized one are shown in Figure 14. Each technique produces a similar amount of amorphous conformations. The decreases in amorphous content by annealing are the same for both systems (Figure 15).

A digital comparison of room-temperature spectra of the rocking mode for slow-crystallized and isopentane-quenched samples is shown in Figure 16. Surface orientation was induced upon quenching. Therefore, the polarization effects preclude digital comparisons of these systems

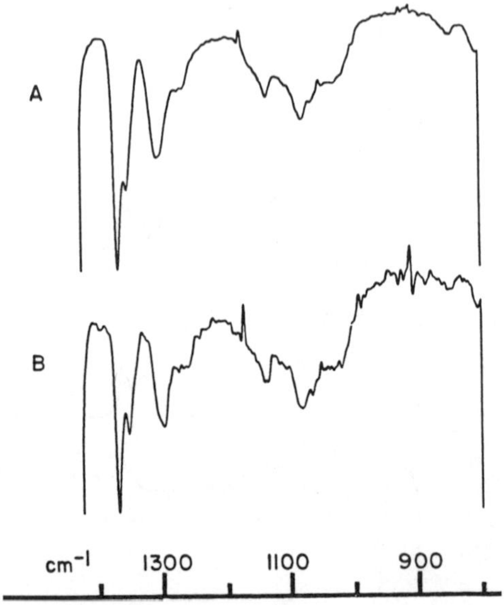

Figure 11. Difference spectra for slow-crystallized minus quenched in isopentane slurry. (A), room-temperature subtraction (1:1.2); (B), 78 K subtraction (1:1.95).

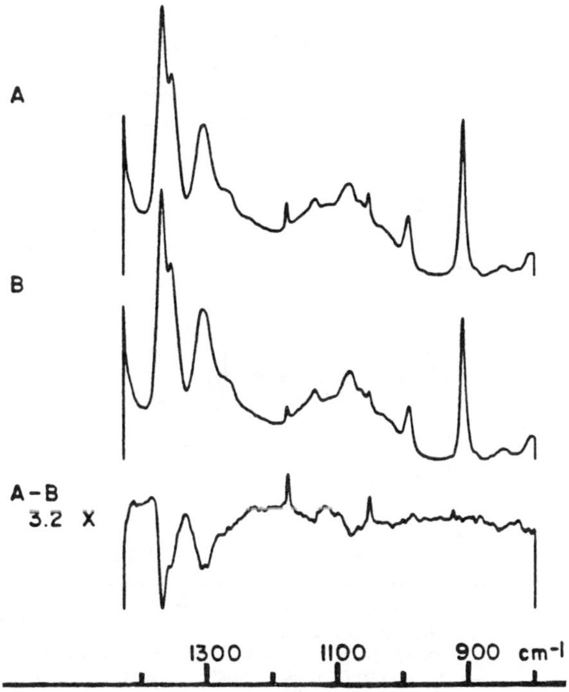

Figure 12. Subtraction of spectra at room temperature. (A), quenched and annealed; (B), quenched in isopentane slurry; (A − B), subtraction (1:0.95).

using the rocking mode. It is only proper to compare quenched systems before and after annealing. In these samples the orientation of the crystallites should not change when annealing below the melt. After annealing a quenched sample, the 731 cm^{-1} band increases in intensity. The subtraction of room-temperature spectra for the isopentane quench shows this effect (Figure 17). The difference peak is at a slightly higher frequency than 731 cm^{-1}, however. This also occurs for the sample quenched in ice water. Figure 18 shows the similarity of the difference spectra for both quenching methods. The subtraction of the spectra of the isopentane-quenched sample from the annealed sample taken at the cryostat's base temperature shows a different result (Figure 19). The increase in intensity of the interaction peak is still caused by a component of higher frequency. However, a decrease of a broad absorption under the 720 cm^{-1} band is observed. The quenched sample had only been annealing at room temperature for 10 hr before observation. In this period crystallization was not completed at room temperature. The spectrum for this sample at room temperature, which was subtracted in Figure 17, was taken after the temperature survey was completed and additional

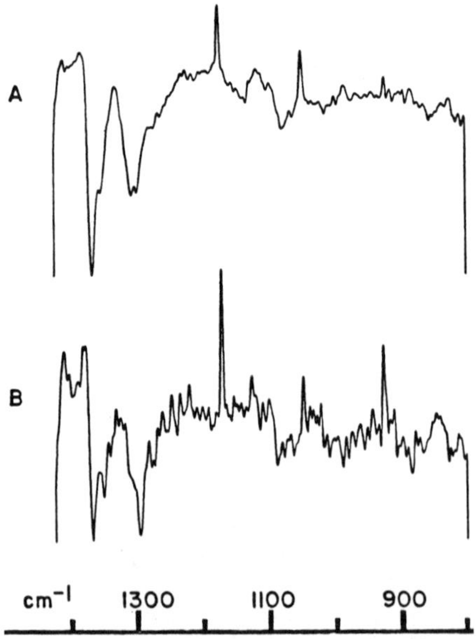

Figure 13. Difference spectra of quenched and annealed minus quenched in isopentane slurry. (A), room temperature (1:0.95); (B), 78 K (1:0.93).

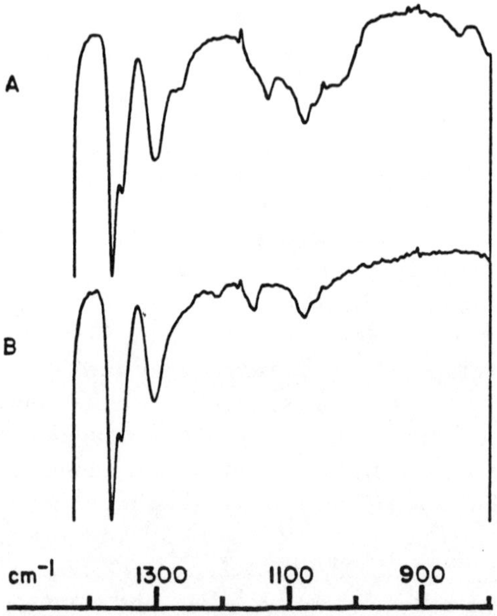

Figure 14. Room-temperature difference spectra of slow-crystallized minus quenched in: (A), isopentane slurry (1:1.2); (B), ice water (1:1.68)

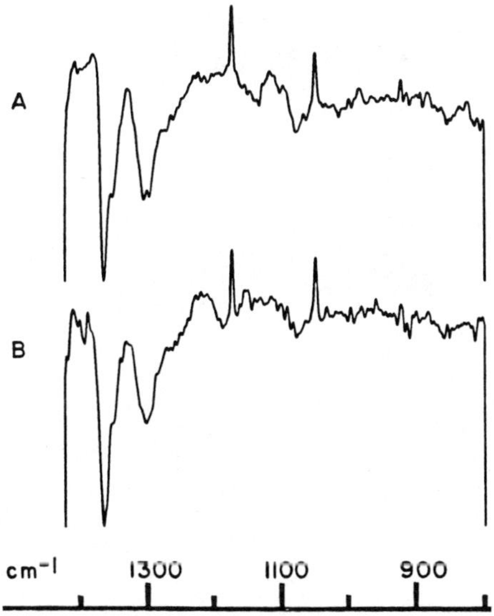

Figure 15. Room-temperature difference spectra of quenched and annealed minus quenched in: (A), isopentane slurry (1:0.95); (B), ice water (1:0.93)

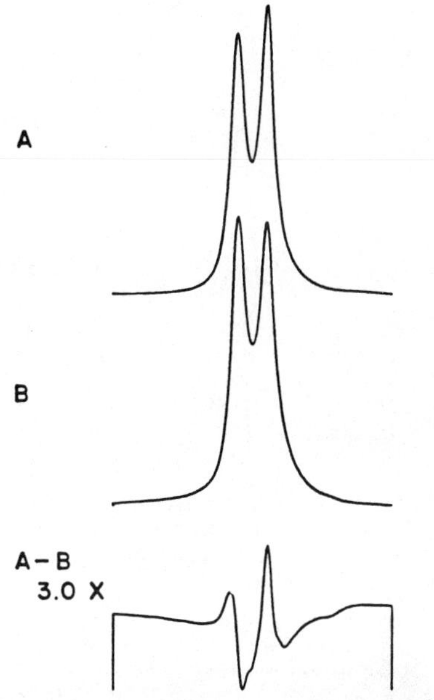

Figure 16. Subtraction of spectra at room temperature. (A), slow-crystallized; (B), quenched in isopentane slurry; (A − B), subtraction (1:1.2).

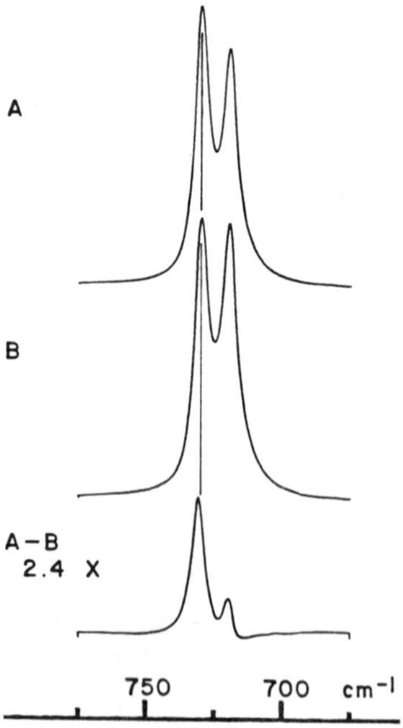

Figure 17. Subtraction of spectra at room temperature. (A), quenched and annealed; (B), quenched in isopentane slurry; (A − B), subtraction (1:0.95).

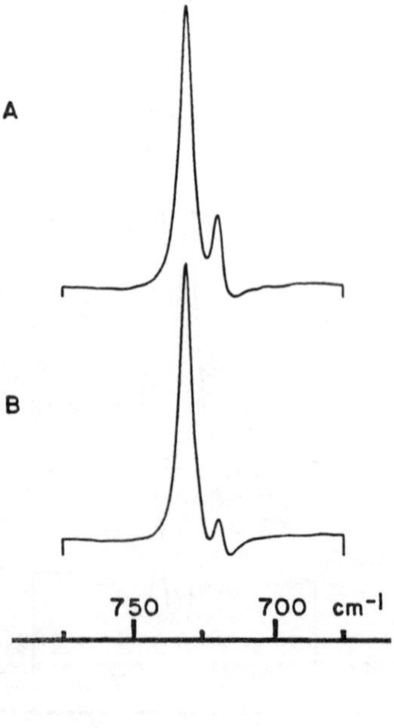

Figure 18. Room-temperature difference spectra of quenched and annealed minus quenched in: (A), isopentane slurry (1:0.95); (B), ice water (1:0.93)

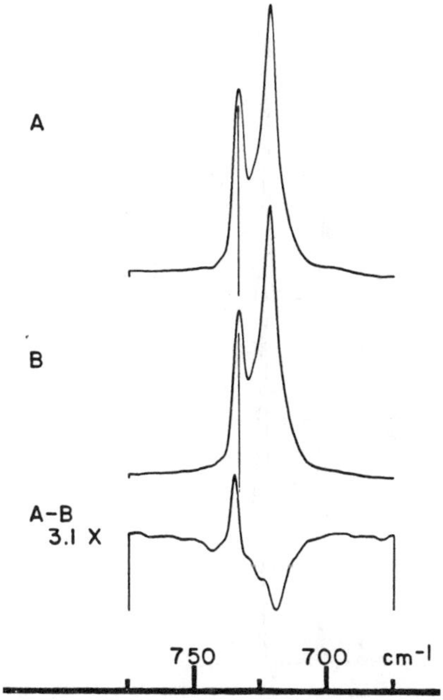

Figure 19. Subtraction of spectra at 78 K. (A), quenched and annealed; (B), quenched in isopentane slurry; (A − B), subtraction (1:0.93).

crystallization had occurred. Therefore, the negative difference peak is attributed to the amorphous component of the 720 cm^{-1} peak which decreased upon further crystallization.

Additional information is available by subtracting spectra taken at different temperatures for the same sample. Since the position of the film was not altered as the temperatures varied, a 1:1 subtraction is a systematic method to illustrate the thermal effects. The spectra at the two temperature extremes for a slow-crystallized sample are subtracted in Figure 20. The 909 and 990 cm^{-1} vinyl bands narrow in width, increase in intensity, and shift to slightly higher frequency as the temperature is decreased. The crystalline absorptions at 1050 and 1176 cm^{-1} shift to lower frequency and sharpen considerably. The 1303 cm^{-1} amorphous absorption shifts its maximum to 1300 cm^{-1} and possibly increases in intensity. The 1353 cm^{-1} band remains fixed in position. The most intense amorphous band, 1369 cm^{-1}, moves to 1371 cm^{-1} at 78 K and has an intensity increase. These results are shown clearly in the difference spectrum. Similar results are obtained for the isopentane-quenched sample before and after annealing.

It is not obvious that the change in frequency of the 1303 and 1369 cm^{-1} absorption band is strictly a shifting of the original peaks. Perhaps a sequence of conformations appears at lower temperatures giving rise to a slightly different frequency than the sequences giving rise to these

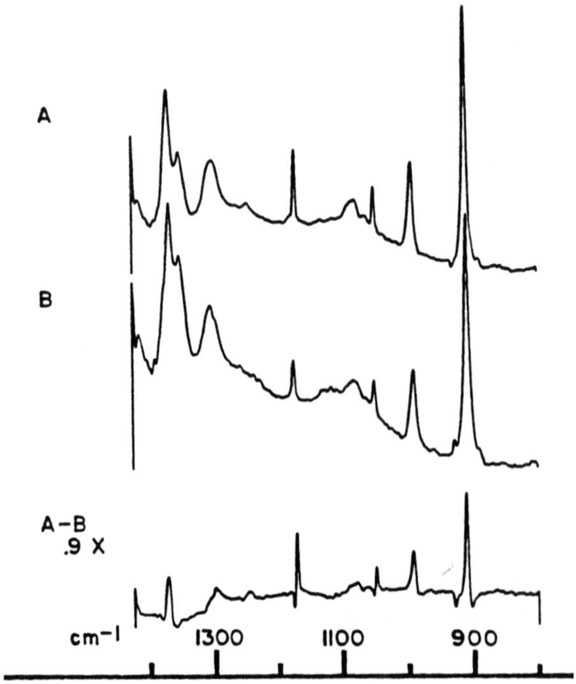

Figure 20. Spectra of slow-crystallized PE in cryostat. (A), 78 K; (B), 295 K; (A − B), subtraction (1:1).

peaks at room temperature. This seems particularly plausible when one observes the narrow difference peak at 1373 cm^{-1}. To further explore this possibility, difference spectra corresponding to other temperatures are presented in Figure 21. The subtractions for below and above T_g are spectra B and D, respectively. Both cases show the difference peak at 1373 cm^{-1}. However, subtraction C for the temperature range including T_g also shows a decrease in the absorption at 1363 cm^{-1}. It is significant that the 1373 cm^{-1} band increases continuously as the temperature is lowered but that the 1368 cm^{-1} absorption only decreases or shifts in the temperature region containing the glass transitions.

The shifting of the rocking mode with decreasing temperature has been investigated thoroughly (*48, 49*). These effects are observed here also. The 720 and 731 cm^{-1} peaks move to 722 and 734 cm^{-1}, respectively, at 78 K. The subtraction of the spectra of a slow-crystallized sample at the two temperature extremes is shown in Figure 22. Any changes in intensities are obscured by the frequency shifts. Similar difference spectra are observed for the quenched and annealed samples.

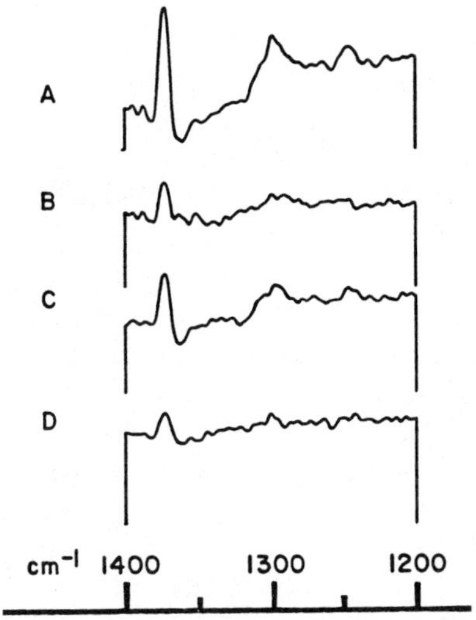

Figure 21. Difference spectra of slow-crystallized sample at different temperatures (1:1). (A), 78–295 K; (B), 78 151 K; (C), 151–259 K; (D), 259–295 K. All subtractions are plotted on the same absorbance scale.

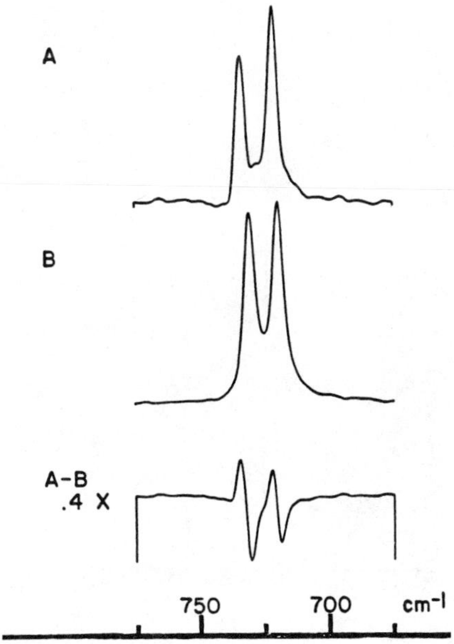

Figure 22. Subtraction of spectra of slow-crystallized sample. (A), 78 K; (B), 295 K; (A − B), subtraction (1:1).

Peak height measurements were performed using baselines which were illustrated in Figures 2 and 3. Interpolation has also been discussed previously. The reproducibility of data acquisition for the FTIR depends on a high signal-to-noise ratio. The use of a cryostat, however, is less than the optimal sampling method. Lower energy throughput reduces the signal-to-noise ratio. Condensation of water and carbon dioxide cause difficulties in obtaining a high quality spectrum. The lowest signal-to-noise ratios of the spectra determined the maximum error of intensity measurement. This upper bound was 3% for the rocking mode and 5% for the absorptions in the region 1425–800 cm^{-1}.

A plot of peak height intensity vs. temperature for the 720 cm^{-1} band is shown in Figure 23. In this figure and all similar ones, the data is normalized to the lowest temperature measurement and is offset for clarity. The slow-crystallized sample shows a continual rise of intensity starting at the γ_{11} relaxation, ca. 110 K. This increase ends at approximately $T_g(U)$ or 240 K. The isopentane-quenched sample, however, remains essentially constant in intensity until $T_g(U)$, whereupon it also shows a decrease. After the quenched sample is annealed, the intensity has minor fluctuations at γ_{11} and in the upper glass transition region.

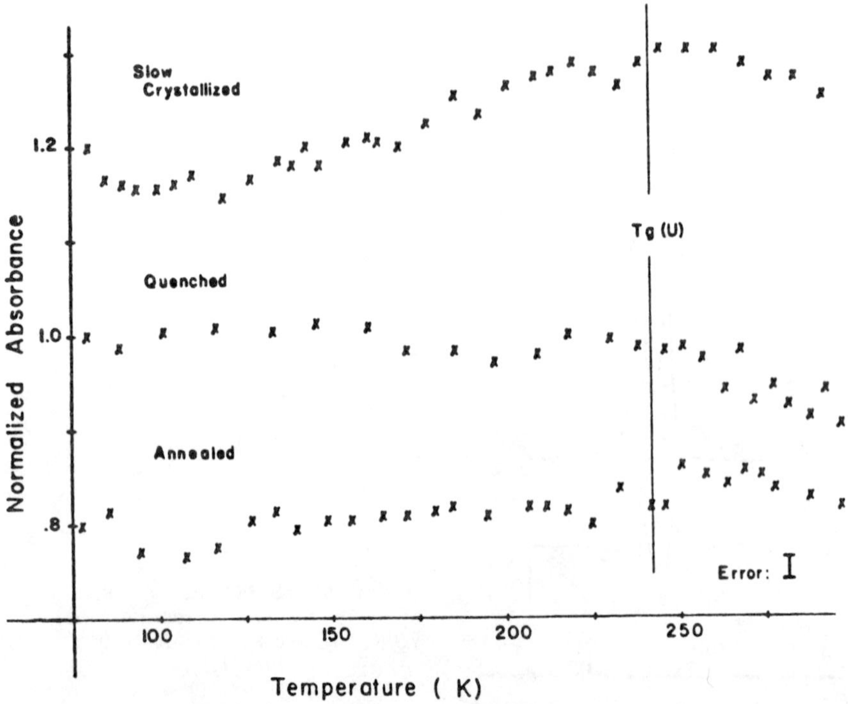

Figure 23. Normalized interpolated peak height of 720 cm^{-1} absorption vs. temperature. Maximum error of measurement, 3%.

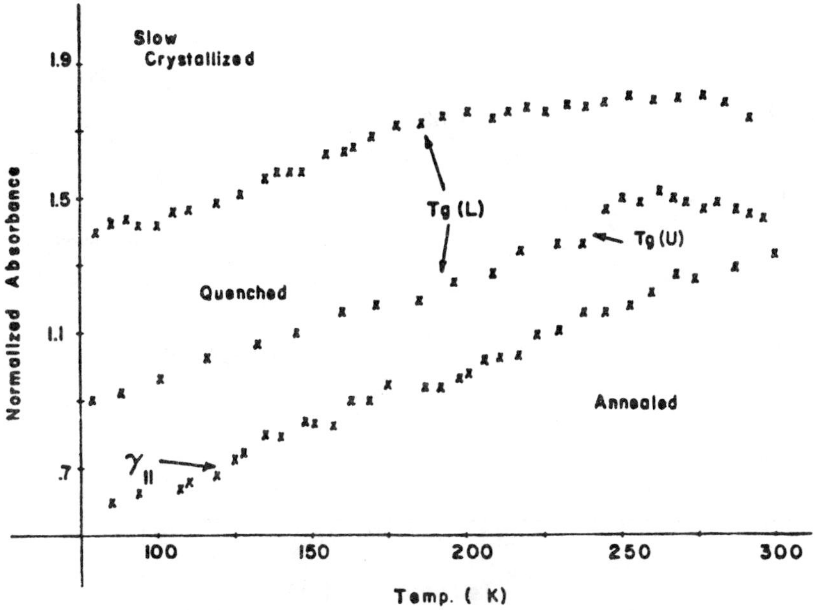

Figure 24. Normalized interpolated peak height of 731 cm^{-1} absorption vs. temperature. Maximum error of measurement, 3% (within data marker).

The changes of the 731 cm^{-1} absorption band with temperatures are much larger and can be seen in Figure 24. The 3% error bar is within the size of the data marker in this case. The intensity for the low-crystallized sample shows the lower glass transition clearly by a change in slope at 190 ± 5 K. Although there may be a minor inflection at T_g(L) for the quenched PE, the major effect is at T_g(U) where an abrupt increase in intensity occurs. After the quenched sample is annealed, the intensity increases gradually with only minor inflections.

The 725 cm^{-1} doublet has no overlapping peaks and a high signal-to-noise ratio. These factors allow a more accurate integration to be performed than is possible with the other absorptions studied. The trapezoid method was used to perform the integration of the digital spectra. Integral intensities of the absorption from 775 to 650 cm^{-1} are presented in Figure 25. The slow-crystallized sample has the largest increase in absorbance with discontinuities at T_g(L) and T_g(U). The absorbance increase does not occur until after the γ_{11} relaxation. After quenching, no gradual rise is observed after γ_{11}. The only absorbance increases occur at specific transitions, which are the γ_1, T_g(L), and T_g(U). Annealing produces a temperature response which has components of both the other morphologies. Again, both glass transitions are observed.

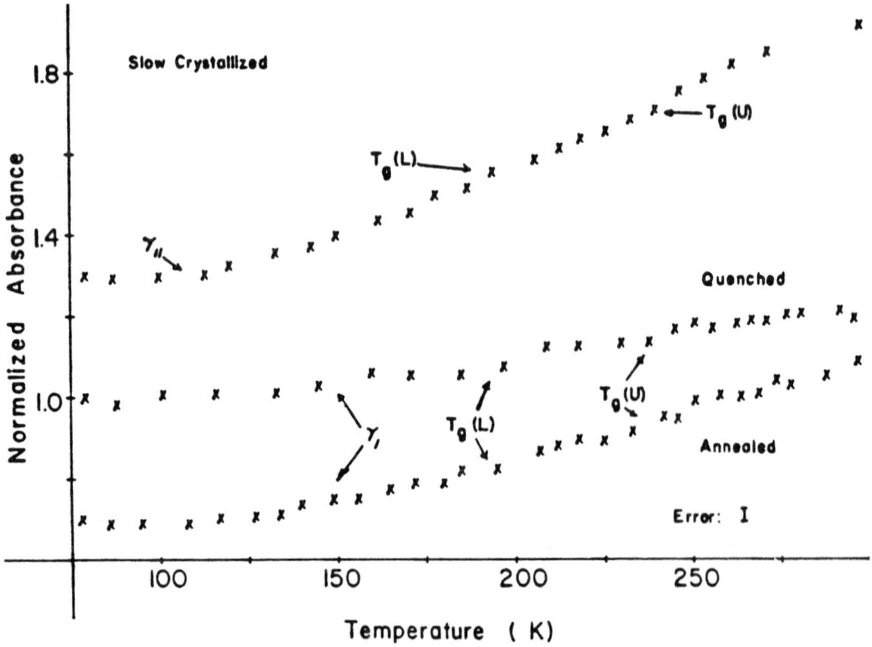

Figure 25. Normalized integrated absorbance of 725 cm⁻¹ doublet vs. temperature. Maximum error of measurement, 3%.

The intensities of two amorphous bands follow similar trends for all the samples studied. The 1303 cm^{-1} absorption is shown in Figure 26 to decrease gradually in intensity until reaching the region of the T_g. Subsequently, the intensity remains constant. This same behavior is seen for the 1369 cm^{-1} band in Figure 27. The intensity of the 1353 cm^{-1} peak is different, as shown in Figure 28. For the quenched and annealed samples, the peak height is constant until $T_g(U)$. At this point it starts to increase. The intensity for the slow-crystallized sample gradually increases throughout the temperature variation. No definite change can be seen at $T_g(U)$ for this morphology.

The 909 cm^{-1} band is fairly isolated in frequency from the other absorptions and would normally be integrated easily. However, as compensation by the reference beam becomes poorer from condensation on the cryostat, the absorption at 926 cm^{-1} fluctuates and imparts large errors to the integration. Therefore, the best monitor that is available is the peak height (Figure 29). Quenched and annealed samples show definite changes in their slopes at $T_g(L)$. The γ_1 and $T_g(U)$ transitions correspond to the temperatures where changes of slope are noticed for the slow-crystallized sample's intensity.

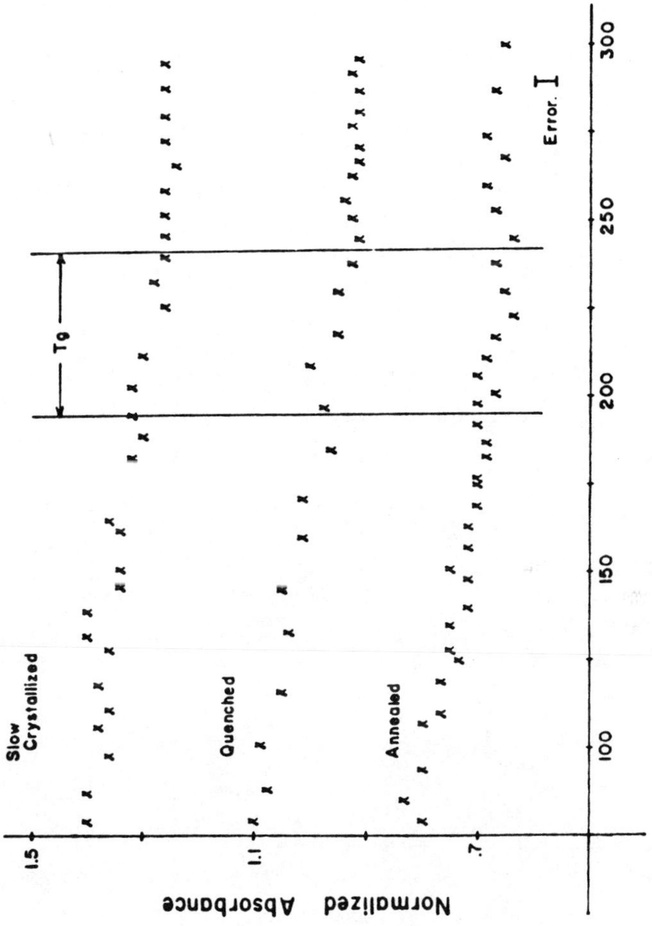

Figure 26. Normalized interpolated peak height of 1303 cm^{-1} absorption vs. temperature. Maximum error of measurement, 5%.

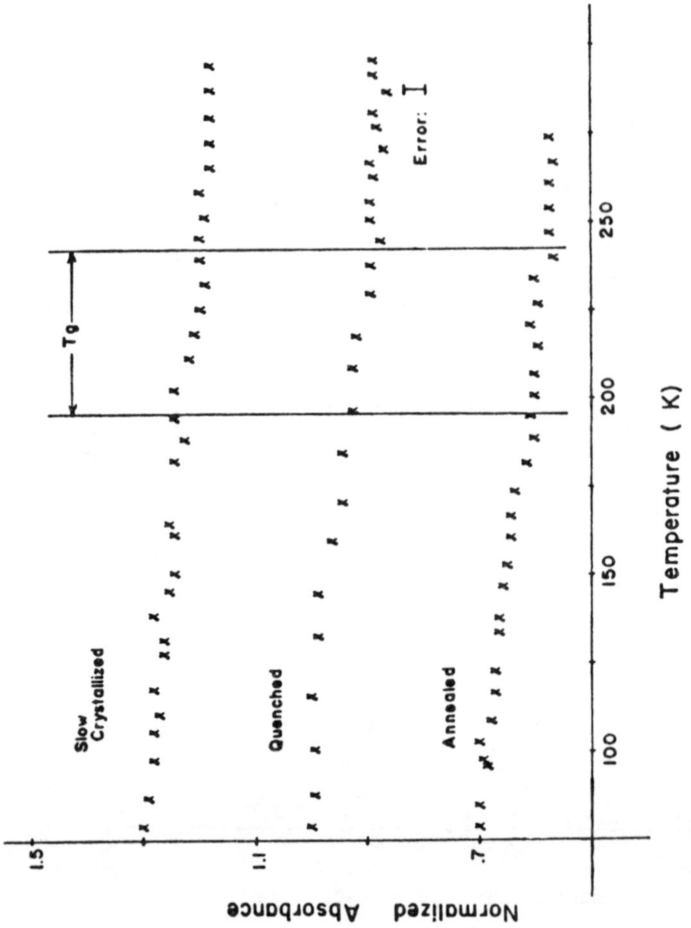

Figure 27. Normalized interpolated peak height of 1369 cm^{-1} absorption vs. temperature. Maximum error of measurement, 5%.

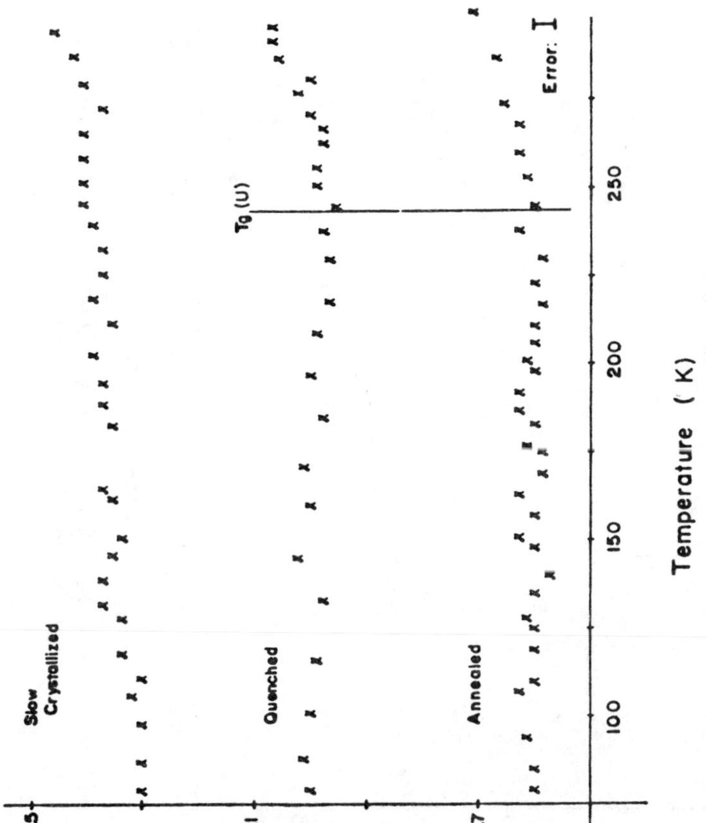

Figure 28. Normalized interpolated peak height of 1353 cm^{-1} absorption vs. temperature. Maximum error of measurement, 5%.

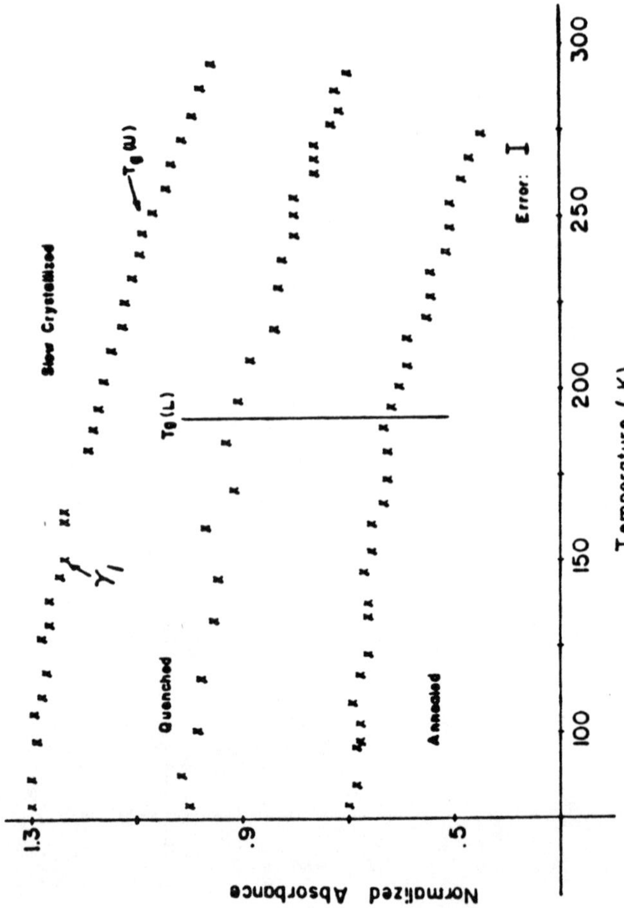

Figure 29. Normalized interpolated peak height of 909 cm^{-1} *absorption vs. temperature. Maximum error of measurement, 5%.*

Analysis of the above results show that certain effects may appear in the measurement of intensities as a consequence of factors which were not experimentally or instrumentally controlled. The band intensities can be affected by sample orientation and nonuniform film thickness. When comparing the spectra of the same sample as a function of temperature, these effects should be minimal. Intensity measurements of the 725 cm^{-1} rocking mode are also susceptible to errors because of its strong absorption. In all samples studied, the maximum absorbance was never greater than 1.2. The heating and cooling rates may introduce other effects. Since a similar time period (15 min) elapsed before the spectrum was recorded at each temperature, the heating rate is assumed to be minimal. Even with these hesitations of consience, it is felt that the present data is superior to that previously available in the literature.

Discussion

The temperature dependences of the IR bands of PE were quite different for the slow-crystallized and quenched samples. This effect was most noticeable for the crystalline rocking modes. The peak heights of the 720 and 731 cm^{-1} components of this doublet increased with increasing temperature. This was not as anticipated. The intermolecular forces lessen with thermal expansion. Therefore, the induced dipole moment of the interaction decreases with increasing temperature (*45, 46*). The intensity of a band which is enhanced by the dipole–dipole interaction is expected to decrease.

A change in peak height can sometimes be attributed to changes in band shape. The width of an absorption normally increases with increasing temperature. This is caused by additional thermal energy being imparted to the vibrational energy levels. The effects of thermal broadening would, therefore, decrease the apparent peak height and cannot explain the anomalous rise of intensity with temperature. Another factor which can be attributed to some of the unexpected intensity increase is the mutual overlapping of the two major components of the rocking doublet. As temperature increases, the splitting decreases from approximately 12 to 11 cm^{-1}. The apparent peak heights would increase because the bands' individual intensities reinforce each other more. Although the small change in splitting may account for some of the effects, it is doubtful whether it can completely describe the increases observed. Curve resolving, which is necessary to separate the multiple components of the rocking mode, was not used in this study. However, integration of the entire doublet should erase any effects of band shape. The integral intensities also increased with temperature. This negates any interpretation of bandwidth changes being totally responsible for the unexpected behavior of peak heights in this case. There are at least two other explanations which can be proposed.

First, let us reconsider the origin of these bands. The rocking mode is split into a doublet by intermolecular interactions resulting from close packing of the two chains in the unit cell (*47*). The 731 and 720 cm^{-1} components are polarized along the *a* and *b* axes, respectively (*62*). Snyder (*47*) has related the relative intensities of the two bands in n-$C_{23}H_{48}$ to the setting angle in the crystal by the equation:

$$\frac{I_a}{I_b} = \tan^2\theta$$

where θ is the angle between the plane containing the skeleton of the chain and the *a* axis. This was done with the assumption that ". . . the intensities of the two components result from the addition and subtraction of the unperturbed dipoles resulting from the vibrations of the two molecules of the subcell . . ." (*47*). The angle obtained by Snyder's measurement of the relative intensities at 93 K was 42°. This agreed well with the x-ray determination of Smith (*65*) at room temperature for this alkane.

The magnitude of the setting angle of PE and similar paraffins and its variance at low temperature are not well known at present. The room-temperature x-ray values range between 42° and 48° (*65, 66, 67, 68*). An angle of 41° was measured by neutron diffraction at 4 and 90 K (*69*). The error of this determination was not published. The uncertainty of x-ray measurements is usually ±6° (*68*). This puts the values within range of one another. X-ray data of Kasai and Kakudo at 173, 291, and 363 K suggested that the setting angle decreases with increasing temperature (*70*). Kavesh and Schultz further supported this conclusion with measurements above room temperature (*68*).

If values of $\theta > 45°$ are used in conjunction with Snyder's equation, then the predicted ratios of the relative intensities are greater than one. The observed ratio for the slow-crystallized sample of the 731–720 cm^{-1} bands was always less than unity at room temperature. Furthermore, the decrease of the setting angle with increasing temperature predicts that the 731 cm^{-1} band would decrease and the 720 cm^{-1} band would increase. This was also not observed. The direct transferrability of this equation to PE is doubtful since the 720 cm^{-1} absorption has components which result from amorphous as well as from crystalline methylene rocking. Even so, the trends with temperature were not as predicted. The integral intensity of the doublet also increased with temperature. These experimental observations imply that the original additivity assumption of Snyder is poor. Therefore, an explanation other than setting angle must be proposed.

Nikitina et al. observed a similar temperature dependence of the intensities of the amide I and II bands in the IR spectra of various nylons (*30*). The basis of these observations was unclear. The amide bands of nylons and the 725 cm^{-1} rocking mode of PE both indicate structural order. The interactions that exist in the unit cell cause the splitting of the rocking mode. Various calculations have been performed demonstrating the effects of these forces on the frequencies of the vibrations (*47, 57, 71, 72, 73*). The usual dipole–dipole interactions are not sufficient for the large splitting observed. This was first analyzed by Stein (*71*). Interatomic forces arising from hydrogen–hydrogen repulsion are the dominant cause of the band splitting. Tasumi and Krimm calculated the effects of both types of interactions and variations of the setting angle on the splitting of this mode and also of the low frequency lattice modes (*73*). The dipole–dipole forces caused the methylene rocking mode to split to a lower frequency. The crystalline interaction band is observed experimentally to be shifted to higher frequency. This indicated that the interatomic forces were even greater than initially thought for the rocking mode. They must cancel the effect of dipole–dipole interactions and still supply the needed potential to attain the experimental 11 cm^{-1} splitting.

The hydrogen–hydrogen repulsions may be strong enough at low temperatures to perturb the rocking vibration. This could subsequently lessen the observed intensity. Several experimental observations support this hypothesis. First, the rise of peak height with increasing temperature was largest for the 731 cm^{-1} band. Thermal expansion of the crystal occurs primarily along the *a* axis (*74*). This is also the direction of the vibrational transition moment for the 731 cm^{-1} absorption. The increase for the 720 cm^{-1} band was much smaller. The corresponding expansion along the *b* axis is an order of magnitude less than for the *a* (*74*). Therefore, thermal expansion relieved the constraints imposed on these vibrations by repulsive forces, and the intensities increased accordingly.

The rate of increase of intensity with temperature for the 720 cm^{-1} band depended on sample preparation. A similar effect was also observed for the integral intensity of the entire rocking doublet. This can be attributed to the original reason for the anomalous rise of intensity. A better understanding of the correlation between these results and the hypothesis can be obtained by recalling the comparisons of spectra taken at room temperature. The amorphous wagging modes showed a twofold increase for the quenched sample compared with the slow-crystallized one. This was in good agreement with the morphology observed by Jones et al. by electron microscopy studies of similarly quenched films (*53, 75*). They observed modules and poorly formed spherulites composed on very small crystallites. The distribution of crystallite sizes and/or the concentration of defects was presumably large since DSC scans gave no well defined melting point. The films used in the present IR studies were much thicker than those of Jones (25–50 μm vs. 1000 Å) and were insulated further by the Teflon-coated aluminum foil during the quench. Therefore, the distortions of a well formed spherulitic structure may not have been as large. Neverthless, the crystallites formed by the rapid quench method are assumed to be very imperfect relative to slow crystallization.

Annealing the quenched sample at 100°C produced only a small decrease in the amorphous bands. Since recrystallization normally occurs at temperatures greater than 120°C, this result was expected. (The texts by Wunderlich contain a more complete discussion of crystallization, recrystallization, annealing, and defect structure (*76*).) The annealing temperature was above the α_c transition, which is usually associated with crystalline motion in the lamellae (*7*). Therefore, defects could be excluded from the crystal at this temperature. In the region of the rocking doublet, the interaction band at 731 cm^{-1} increased in intensity. When the spectra were subtracted, the increase of this band was seen to result from a component at slightly higher frequency than the initial 731 cm^{-1} absorption. This implied that the interactions had become stronger when the crystallites had improved their perfection upon annealing.

The samples can be ranked according to quality of crystallite packing as slow-crystallized > quenched and annealed > quenched. This is also the order of the magnitude of slopes observed for the integrated intensities and 720 cm^{-1} peak heights. This gives further support to the previous hypothesis for the unusual intensity decrease with decreasing temperature. When crystalline perfection is improved, the interactions between rocking vibrations increase to a greater extent with thermal contraction. Therefore, the resultant decrease of intensity is more pronounced.

In all observations of annealed and slow-crystallized PE's the rise of intensity with increasing temperature was initiated at approximately 100–110 K. This corresponds well with the γ_{11} transition which has been assigned to local crystalline motion (*5,7*). This result follows since small motions in the crystal relieve the intermolecular forces which are perturbing the vibration, and a subsequent rise of intensity is initiated. The end of this rise in slope and a slight decrease was observed above approximately 250 K for the 720 cm^{-1} peak height. This was consistent for all samples. The integral intensity of the rocking doublet did not reveal this trend. Therefore, it was probably caused by a change in band shape, i.e., a broadening. The 720 cm^{-1} band is comprised of an amorphous and a crystalline component. The limiting frequency for the all trans conformation occurs at 720 cm^{-1}. In actuality, this is approached after a sequence of six methylene units (*54*). Therefore, the amorphous component is a broad peak underlying the large crystalline rocking peak. As the ratio of the gauche to trans conformations increases, this peak broadens asymmetrically to higher frequency (*55*). This would reduce the measured peak height. The observed decrease in intensity above 240 K for the 720 cm^{-1} band could be attributed to changes in the distribution of conformations in the amorphous phase. This agrees well with the volume relaxation measurements showing T_g at 243 K by Davis et al. (*3*).

The peak heights of the 731 cm^{-1} band and the integral intensities of the 725 cm^{-1} doublet observed changes at other temperatures corresponding to amorphous relaxations. This could be interpreted as strains placed upon the crystal by the surrounding amorphous matrix. These forces could be transferred by surface adsorption or by molecular interconnections of the two phases. Additional freedom of molecular motion is transferred to the crystal when the amorphous phase relaxes. The T_g(U) and the T_g(L) were revealed as step increases in the overall background of the integral intensities for all samples. The peak height of the 731 cm^{-1} band was more selective in observing transitions. Rather than step increases, a decrease in slope was observed at 185–190 K for the slow-crystallized sample. This would imply that the phases are well separated and that the major effect seen is from the true amorphous motions of T_g(L). The quenched sample had a larger defect concentration. This sample revealed the T_g(U) as a definite jump in intensity showing that the strains relieved by the amorphous relaxation were severe and probably caused by molecular interconnections of the phases. The interpretation of these changes is in a large part based on the reasoning of Boyer (*2, 4*). The results seem to fit this mechanism quite well. After the quenched sample was annealed, no transition was observed with this crystalline absorption at temperatures corresponding to T_g(U) or T_g(L). This is attributed to the exclusion of entrapped chain ends which were providing most of the interaction of the two phases before annealing. However, the 731 cm^{-1} band is presumably most sensitive to band broadening from the variation of intermolecular forces and subsequent shifting of the peak. This could be obscuring the true nature of these relaxations.

The amorphous methylene wagging bands show a temperature behavior which is more amenable to discussion. The 1352 cm^{-1} band has been calculated to result from the deformation of the methylene isolated by the GG' conformation (*55, 56*). The intensity of this band increased at elevated temperatures relative to the other amorphous wagging modes as a consequence of the higher energy of its conformation (*55, 56*). This was also observed in the present work for the slow-crystallized sample. However, increases for the rapidly quenched systems only occurred after

240 K. This phenomenon can be explained if the rapid quenching produced greater molecular interconnections of the two phases. Tie molecules and loose loops would add an extra hindrance to the formation of the high energy conformers below T_g.

The assignments of the 1369 and 1303 cm^{-1} bands are not as well defined. Snyder attributed these to the symmetric and antisymemtric wagging of methylenes in the GTG structure (*55*). The symmetry arises from the relationship of the deformations of the two methylenes isolated in the GT conformers with respect to the mutual trans bond. A GTTG conformation was shown to possibly be responsible for the 1369 cm^{-1} band by the calculations of Zerbi et al. (*56*). The temperature dependences of the peak heights of both the 1369 and 1303 cm^{-1} bands were very similar in the present study. A slight decrease of intensity occurred with increasing temperature up to the range between T_g(L) and T_g(U). It is not clear whether this is caused by the bands' inherent temperature dependences or by a concentration change; it is probably a combination of the two. Subtractions of spectra at various temperatures showed that the 1369 and 1303 cm^{-1} maxima shifted to higher and lower frequencies, respectively, as the temperature was lowered. If the temperature dependence is strictly thermal in nature, then the absorptions should shift in the same directions. However, the relative spacing of the two peaks may separate if a slightly different conformation is being formed as the temperature is lowered. The latter is thought to be the major factor since a relatively narrow difference band at 1373 cm^{-1} was seen to gradually increase. Yet, this interpretation is very speculative. The only concrete conclusion that can be drawn from these bands is that there is a change in their respective temperature dependencs between T_g(L) and T_g(U).

The absorption at 909 cm^{-1} is attributed to vinyl end group C–H deformations. The intensity decreased with increasing temperature. Inflections in the slope were observed at temperatures corresponding to the T_g(U) and γ_1 transitions for the slow-crystallized sample. After the rapid quenching and subsequent annealing treatments, a gradual change of slope was only discernible at 185–190 K, i.e., T_g(L). This appears to be opposite to the results obtained from the other band intensities. However, the nature of the 909 cm^{-1} absorption must be considered. Since it is associated with vinyl end groups, there is no contribution from main chain conformational changes. Nitric acid oxidation studies of Keller and Priest (*78*) on solution-crystallized polyethylene showed that 90% of these terminal groups were excluded from the crystal. Recent bromination studies by Arroyo (*79*) also support this conclusion for melt-crystallized polyethylene. Therefore, the 909 cm^{-1} band should be a probe of the amorphous phase with only minor effects of chain end inclusion in the crystal. The intensity of amorphous bands was shown to be less for the slow-crystallized sample. This implied greater and more perfect crystallinity than for the quenched systems. The presence of two observed glass transitions was attributed by Boyer (*2*) to variations in the mobility of amorphous molecules which are affected by the degree of crystallinity. This result agrees with that interpretation. However, it also disagrees with the previous results on the 731 cm^{-1} band which were explained in terms of defects. One explanation is that the transitions observed with the 731 cm^{-1} peak heights were influenced strongly by band shape effects. A more speculative interpretation is that the locations of the vinyl groups, although mostly in the amorphous phase, could depend on sample preparation. Perhaps the vinyl groups of the slow-crystallized sample are predominantly on short cilia which are experiencing a larger intimacy with the crystal. This would explain the observation of $T_g(U)$.

Literature Cited

1. Stehling, F. C., Mandelkern, L., *Macromolecules* (1970) **3**, 242.
2. Boyer, R. F., *Macromolecules* (1973) **6**, 288.
3. Davis, G. T., Eby, R. K., *J. Appl. Phys.* (1973) **44**, 4274.
4. Boyer, R. F., *J. Polym. Sci., Polym. Symp.* (1975) **50**, 189.
5. Illers, K. H., *Kolloid Z. Z. Polym.* (1969) **231**, 622.
6. Ibid. (1972) **250**, 426.
7. Ibid. (1973) **251**, 394.
8. Lee, S., Simha, R., *Macromolecules* (1974) **7**, 909.
9. Sanui, K., MacKnight, W. J., Lenz, R. W., *Macromolecules* (1974) **7**, 101.
10. Bergmann, K., Nawotki, K., *Kolloid Z. Z. Polym.* (1967) **219**, 132.
11. Chang, S., *J. Polym. Sci., Polym. Symp.* (1973) **43**, 43.
12. Cooper, J. W., McCrum, N. G., *J. Mater. Sci.* (1972) **7**, 1221.
13. Paauwe, N. A., Knox, J. R., *Polym. Eng. Sci.* (1976) **16**, 36
14. Bergmann, K., *Kolloid Z. Z. Polym.* (1973) **251**, 962.
15. Slichter, W. P., "NMR: Basic Principles and Progress," P. Diehl, E. Fluck, and R. Kosfeld, Eds., Vol. 4, pp. 209–231, Springer-Verlag, New York, 1971.
16. McCall, D. W., "Molecular Dynamics and Structure of Solids," R. S. Carter and J. J. Rush, Eds., N.B.S. Special Publication **301**, pp. 475–537, Washington, DC, 1969.
17. McBrierty, V. J., *Polymer* (1974) **15**, 503.
18. Buchachenko, A. L., Kovarskii, A. L., Vasserman, A. M., "Advances in Polymer Science," Z. A. Rogovin, Ed., pp. 26–57, Wiley, New York, 1974.
19. Kumler, P. L., Boyer, R. F., *Macromolecules* (1976) **9**, 903.
20. Roe, J. M., Simha, R., *Int. J. Polym. Mater.* (1974) **3**, 193.
21. Fukawa, K., *Kogyo Kagaku Zasshi* (1963) **66**, 1605.
22. Havriliak, S., Jr., Roman, N., *Polymer* (1966) **7**, 387.
23. Trapeznikova, O. N., Belopolskaya, T. V., *Vysokomol. Soedin., Ser. A* (1967) **9**, 2659.
24. Araki, Y., *J. Appl. Polym. Sci.* (1967) **11**, 953.
25. Anton, A., *J. Appl. Polym. Sci.* (1968) **12**, 2117.
26. MacKnight, W. J., McKenna, L. W., Read, B. E., Stein, R. S., *J. Phys. Chem.* (1968) **72**, 1122.
27. Alter, H., Hsaio, H. Y., *J. Polym. Sci., Polym. Lett. Ed.* (1968) **6**, 363.
28. Hannon, M. J., Koenig, J. L., *J. Polym. Sci., Part A-2* (1969) **7**, 1085.

29. Bessler, V. E., Bier, G., *Makromol. Chem.* (1969) **122**, 30.
30. Nikitina, O. A., Slovokhotova, N. A., Kargin, V. A., *Vysokomol. Soedin., Ser. A* (1970) **12**, 1514.
31. Tsuge, M., Tanimoto, S., Tanaka, S., *Kogyo Kagaku Zasshi* (1970) **73**, 440.
32. Belopolskaya, T. V., Trapeznikova, O. N., *Vysokomol. Soedin., Ser. A* (1971) **13**, 1119.
33. Ogura, K., Kawamura, S., Sobue, H., *Macromolecules* (1971) **4**, 79.
34. Huang, Y. S., Koenig, J. L., *J. Appl. Polym. Sci.* (1971) **15**, 1237.
35. Ogura, K., *Polym. J.* (1972) **3**, 153.
36. Sakai, K., Sobue, H., *J. Appl. Polym. Sci.* (1972) **16**, 2657.
37. Seymour, R. W., Cooper, S. L., *Macromolecules* (1973) **6**, 48.
38. Ogura, K., Sobue, H., Nakamura, S., *J. Polym. Sci., Polym. Phys. Ed.* (1973) **11**, 2079.
39. White, J. J., Hasan, A. U., Sears, W. C., *J. Macromol. Sci., Phys.* (1973) **B7**, 177.
40. Schmidt, P., Schneider, B., Dirlikov, S., Mihailov, M., *Eur. Polym. J.* (1975) **11**, 229.
41. Schroeder, L. R., Cooper, S. L., *J. Appl. Phys.* (1976) **47**, 4310.
42. Enns, J. B., Boyer, R. F., Ishida, H., Koenig, J. L., unpublished data.
43. Ovander, L. N., *Opt. Spektrosk.* (1961) **11**, 68.
44. Ibid. (1962) **12**, 401.
45. Lisitsa, M. P., Tsyashchenko, Y. P., *Opt. Spektrosk.* (1960) **9**, 99.
46. Ibid., 229.
47. Snyder, R. G., *J. Mol. Spectrosc.* (1961) **7**, 116.
48. Brockmeier, N. F., *J. Appl. Polym. Sci.* (1968) **12**, 2129.
49. Shen, M., Hansen, W. N., Romo, P. C., *J. Chem. Phys.* (1969) **51**, 425.
50. Frank, W., Schmidt, H., Wulff, W., *J. Polym. Sci., Polym. Symp.*, in press.
51. Hendra, P. J., Jobic, H. P., Holland-Moritz, K., *J. Polym. Sci., Polym. Lett. Ed.* (1975) **13**, 365.
52. Boyer, R. F., Snyder, R. G., *J. Polym. Sci., Polym. Lett. Ed.* (1977) **15**, 315.
53. Jones, J. B., Barenberg, S., Geil, P. H., *J. Macromol. Sci., Phys.*, in press.
54. Barnes, J., Fanconi, B., in press.
55. Snyder, R. G., *J. Chem. Phys.* (1967) **47**, 1316.
56. Zerbi, G., Piseri, L., Cabassi, F., *Mol. Phys.* (1971) **22**, 241.
57. Snyder, R. G., Schachtschneider, J. H., *Spectrochim. Acta* (1963) **19**, 85.
58. Koenig, J. L., *Appl. Spectrosc.* (1975) **29**, 293.
59. D'Esposito, L., Koenig, J. L., "Fourier Transform IR: Applications to Chemical Systems," J. R. Ferraro, Ed., Academic, New York, in press.
60. Bell, R. J., "Introductory Fourier Transform Spectroscopy," Academic, New York, 1972.
61. Griffiths, P. R., "Chemical Infrared Fourier Transform Spectroscopy," Wiley, New York, 1975.
62. Krimm, S., *Fortschr. Hochpolym. Forsch.* (1960) **2**, 51.
63. Angood, A. C., Koenig, J. L., *Macromolecules* (1969) **2**, 37.
64. Schonhorn, H., *Macromolecules* (1968) **1**, 145.
65. Smith, A. E., *J. Chem. Phys.* (1953) **21**, 2229.
66. Bunn, C. W., *Trans. Faraday Soc.* (1939) **35**, 482.
67. Teare, P. W., *Acta Crystallogr.* (1959) **12**, 294.
68. Kavesh and Schultz, *J. Polym. Sci., Part A-2* (1970) **8**, 243.
69. Avitabile, G., Napolitano, R., Pirozzi, B., Rouse, K. D., Thomas, M. W., Willis, B. T. M., *J. Polym. Sci., Polym. Lett. Ed.* (1975) **13**, 351.
70. Kasai, N., Kakudo, M., *Rep. Polym. Phys. Jpn.* (1968) **11**, 145.
71. Stein, R. S., *J. Chem. Phys.* (1955) **23**, 734.
72. Tasumi, M., Shimanouchi, T., *J. Chem. Phys.* (1965) **43**, 1245.
73. Tasumi, M., Krimm, S., *J. Chem. Phys.* (1967) **46**, 755.
74. Davis, G. T., Eby, R. K., Colson, J. P., *J. Appl. Phys.* (1970) **41**, 4216.

75. Jones, J. B., Ph.D. Dissertation, Case Western Reserve University, Cleveland, Ohio (1977).
76. Wunderlich, B., "Macromolecular Physics," Vol. 1, Academic, New York, 1973.
77. Ibid., Vol. 2, 1976.
78. Keller, A., Priest, D. J., *J. Macromol. Sci., Phys.* (1968) **B2**, 479.
79. Arroyo, N., M.S. Dissertation, Case Western Reserve University, Cleveland, Ohio (1977).

RECEIVED January 12, 1978.

9

Brillouin Scattering and Polymer Science

G. D. PATTERSON

Bell Laboratories, Murray Hill, NJ 07974

Brillouin scattering measures the velocity and attenuation of hypersonic thermal acoustic phonons using light scattering. This technique has been applied now to many problems in polymer science. This chapter describes the theory and experimental procedures used in Brillouin scattering. Many examples are presented of the types of information that can be obtained. These include the adiabatic and isothermal compressibility, the volume and shear viscosity, and the ratio of specific heats for low-viscosity fluids. When the viscosity increases, the hypersonic glass–rubber relaxation is observed and it correlates well with other dynamic mechanical and dielectric data. Near the glass transition one can obtain the high-frequency limiting moduli for compression and shear, the Poisson ratio; and from measurements of the attenuation one can determine the homogeneity of the sample.

Light scattering in dense media is caused by fluctuations in the local dielectric tensor ϵ (*1*). In 1922 Brillouin (*2*) predicted that thermal acoustic phonons would lead to such fluctuations and hence to light scattering. In addition, the scattered light should be shifted in frequency because the phonons are moving. The frequency shift is given by:

$$\pm\Delta\omega/\omega_o = 2n(V/C)\sin\theta/2 \qquad (1)$$

where ω_o is the incident frequency, n is the refractive index, V is the phonon velocity, C is the speed of light in a vacuum, and θ is the scattering angle in the scattering plane. Because the acoustic phonons in fluids and amorphous solids are damped, the shifted Brillouin peaks have a half-width at half-height given by (*3*):

$$\Gamma = \frac{\alpha V}{2\pi} \qquad (2)$$

0-8412-0406-3/79/33-174-141$05.50/1

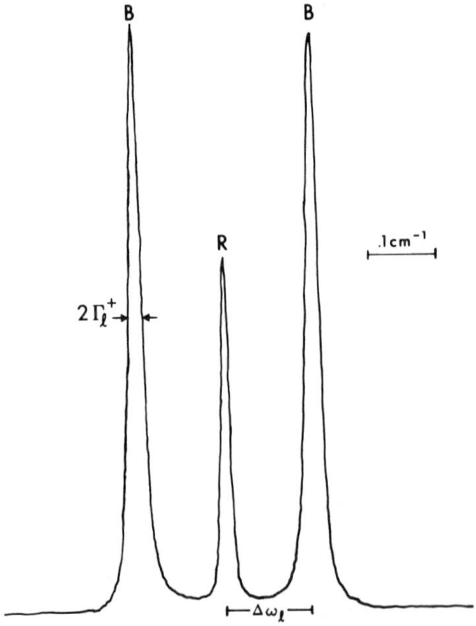

Figure 1. Rayleigh–Brillouin Spectrum of n-*hexadecane at 120°C. Only the central peak caused by thermal expansion is clearly shown in this spectrum.*

where α is the phonon attenuation coefficient and Γ is measured in Hertz. In addition there will be several unshifted peaks with finite width because of thermal expansion, mechanical and thermal relaxation, and optical anisotropy fluctuations. The Rayleigh–Brillouin spectrum of *n*-hexadecane at 120°C is shown in Figure 1.

In this chapter we will describe the theory and experimental techniques of Brillouin scattering. Applications will be made to several polymeric systems and the type of information that can be obtained will be discussed.

Theory

The propagation of acoustic phonons in amorphous media depends on the mechanical and thermal moduli. We will denote the modulus of compression by K, the shear modulus by G, the longitudinal modulus by $M = K + 4/3G$, the thermal conductivity by κ, the thermal expansion coefficient by α, and the ratio of specific heats by $\gamma = C_P/C_V$.

The dispersion equation for longitudinal phonons is:

$$(Mq^2 - \rho\omega^2)\left(\frac{\rho C_V}{T} + \frac{\kappa q^2}{i\omega T}\right) + K^2\alpha^2 q^2 = 0 \quad (3)$$

where $q = 2\pi/\lambda_P$ is the magnitude of the propagation vector for phonons with wavelength λ_P. Brillouin scattering is caused by the creation or annihilation of an acoustic phonon. The value of q is determined by the scattering angle and the wavelength of the light in the medium:

$$q = \frac{4\pi n}{\lambda} \sin \theta/2 \quad (4)$$

where λ is the wavelength of the light in a vacuum. The longitudinal Brillouin splitting is given by the real part of the two complex roots of Equation 3 and the linewidth by the imaginary part of the complex roots. Pure imaginary roots correspond to peaks centered at the incident frequency. The dynamic mechanical loss associated with the longitudinal waves is given by:

$$\tan \delta = \frac{2\Gamma_{(1)}\Delta\omega_{(1)}}{\Delta\omega_{(1)}{}^2 - \Gamma_{(1)}{}^2} \quad (5)$$

The corresponding dispersion equation for the transverse phonons is:

$$Gq^2 - \rho\omega^2 = 0 \quad (6)$$

The spectrum of amorphous bisphenol-A polycarbonate showing both longitudinal and transverse Brillouin peaks is shown in Figure 2.

The solutions to Equations 3 and 6 depend on the form adopted for the moduli. The moduli are in general complex and can be represented as:

$$M(\omega,p,T) = M'(\omega,p,T) + iM''(\omega,p,T) \quad (7)$$

where M′ and M″ are the real and imaginary parts, respectively. The results appropriate to each relaxation time regime will be given in the discussion section.

Typical Brillouin splittings are in the range 10^8–10^{10} Hz. The frequency of the acoustic phonons being studied is given directly by $\Delta\omega$. This can be seen easily from the relation for the velocity:

$$\Delta\omega = qV = \omega_p \quad (8)$$

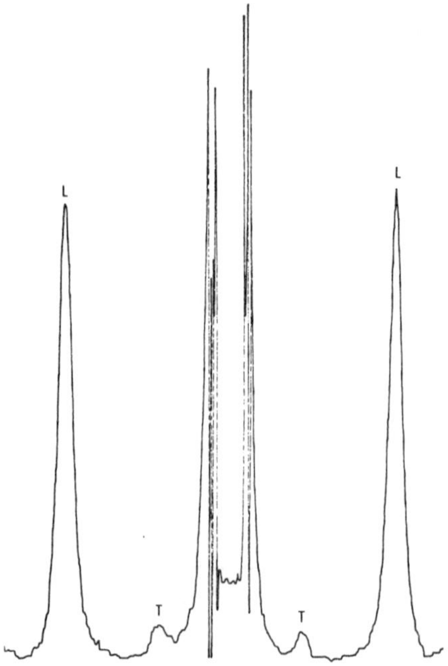

Figure 2. Rayleigh–Brillouin spectrum of amorphous bisphenol–A polycarbonate at 25°C showing both longitudinal (L) and transverse (T) peaks

where ω_p is the phonon frequency. Although high-resolution optical spectroscopy is used as the tool, Brillouin scattering should be considered a dynamic mechanical technique in the hypersonic frequency range.

Experimental

The apparatus necessary to measure the Brillouin spectrum of polymers is now well developed. The first requirement is an intense source of collimated light with a narrow frequency distribution. An argon ion laser provides a bright incident beam. If the laser has a cavity length of 1 m, the output will consist of a series of lines separated by 150 MHz. A single cavity mode may be selected by introducing an etalon into the laser cavity with a free spectral range of 10 GHz. The resulting single line has a width of approximately 10 MHz. Over 1 W has been obtained under these conditions for the 5145-Å line.

The longitudinal Brillouin spectrum appears when both the incident and scattered light are polarized vertically with respect to the scattering plane (I_{VV}). The transverse Brillouin spectrum results when the incident polarization is perpendicular to the scattered polarization ($I_{HV} = I_{VH}$).

Thus, the polarization of both the incident and scattered light must be controlled. A typical argon laser produces linearly polarized light. The incident polarization can be adjusted to be either vertical or horizontal with respect to the scattering plane with a double Fresnel rhomb polarization rotator. The scattered light can be analyzed with a Glan–Thompson prism polarizer.

In order to increase the power density in the scattering volume, the incident light can be focussed. Beam diameters of 100–200 μ are optimal for most work.

A variety of sample configurations is possible. For most amorphous samples above the glass transition a square quartz cuvette is convenient for observation at 90°. If the sample is to be cooled below $T_{(g)}$, a round test tube will withstand more strain and resist cracking. Many polymer samples can be prepared most easily as films. As long as the film is thick enough to have a measurable scattering volume, Brillouin spectra may be obtained. Successful spectra also have been collected from tensile bars, extruded rods, and other common mechanical specimens. The main requirement is that the scattering geometry be known and that the incident and scattered beams be well defined. The scattered light is collimated and analyzed for polarization.

The instrument most commonly used to resolve the Brillouin spectrum is the Fabry–Perot interferometer. This device consists of a pair of highly reflective, optically polished mirrors. The transmission function for a plane parallel Fabry–Perot interferometer is:

$$\frac{I(\omega)}{I_o} = \frac{(T^2/(1-R)^2)}{1 + \frac{4F^2}{\pi^2}\sin^2\left(\omega\frac{nd}{C}\right)} \tag{9}$$

where T is the transmission of the mirrors, R is the reflectivity, F is called the finesse, n is the refractive index of the medium between the plates, d is the distance between the plates, and C is the speed of light in a vacuum. When $\omega = (\pi CN/nd)$, where N is an integer, the interferometer will have its maximum transmission:

$$\frac{I_{max}}{I_o} = \left(\frac{T}{1-R}\right)^2 = \left(1 - \frac{A}{1-R}\right)^2 \tag{10}$$

where A is the absorption of the mirrors. With modern dielectric coatings, the maximum transmission can be quite high.

The spacing between maxima is equal to $\pi C/nd$ and is called the free spectral range. The finesse F is equal to the free spectral range divided by the full width at half-height of the transmission function. The finesse is determined by many factors, but the two most important variables are the reflectivity and surface flatness of the mirrors. The reflectivity finesse is given by:

$$F_R = \frac{\pi \sqrt{R}}{1 - R} \tag{11}$$

and the flatness finesse by:

$$F_F = \frac{M}{2} \tag{12}$$

where M is from the surface flatness expressed as λ/M. The total finesse F is then determined from:

$$\frac{1}{F^2} = \sum_i \frac{1}{F_i^2} \tag{13}$$

Total finesses of the order of 50 can be obtained routinely for $\lambda/200$ plates with $R = 98\%$. Higher reflectivities only lower the transmission without appreciably increasing the finesse because other factors become dominant.

The other important characteristic of a Fabry–Perot interferometer is the contrast, $C \approx 4F^2/\pi^2$. Typical contrasts are near 1000. However, for many bulk polymer samples partial crystallinity, particulate impurities, or anisotropic light scattering may lead to unshifted peak intensities that are 10^4 to 10^7 times greater than the Brillouin intensities. The Brillouin peaks will be hidden then under the very intense wings of the central peak. The problem of contrast can be overcome by the use of the multipass interferometer (*4*). When the light is passed through the interferometer more than one time, the composite transmission function is the product of the single-pass transmission functions. A typical system uses three passes. The interferometer used in this laboratory (*5*) uses $\lambda/200$ plates with $R = 90\%$. The triple-pass finesse is given by:

$$F_3 = \frac{F_1}{\sqrt{2^{1/3} - 1}} \tag{14}$$

so that the observed finesse is 60 and the contrast is greater than 10^7. The above conditions are sufficient for most work. Greater contrast can be achieved by using five passes but the alignment is much more difficult.

The interferometer can be scanned by changing the optical path length between the plates with a piezoelectric transducer. The resolving element is a pinhole which is imaged on the scattering volume. For most work, the pinhole size can be chosen to maximize transmission. A 200-μ pinhole is useful unless the free spectral range is less than about 5 GHz. For higher resolution work, a smaller pinhole must be used to maintain high finesse.

Because the Fabry–Perot interferometer is a comb filter, an additional narrow bandpass interference filter is necessary to isolate the Brillouin scattering and reject Raman scattering or fluorescence.

The light is detected with a photomultiplier and a photon counter. The spectrum is recorded with a multichannel analyzer and displayed as a digital spectrum on an oscilloscope. The observed spectrum can be analyzed directly or if necessary the digital spectrum can be fit with a computer. A diagram of the spectrometer is shown in Figure 3.

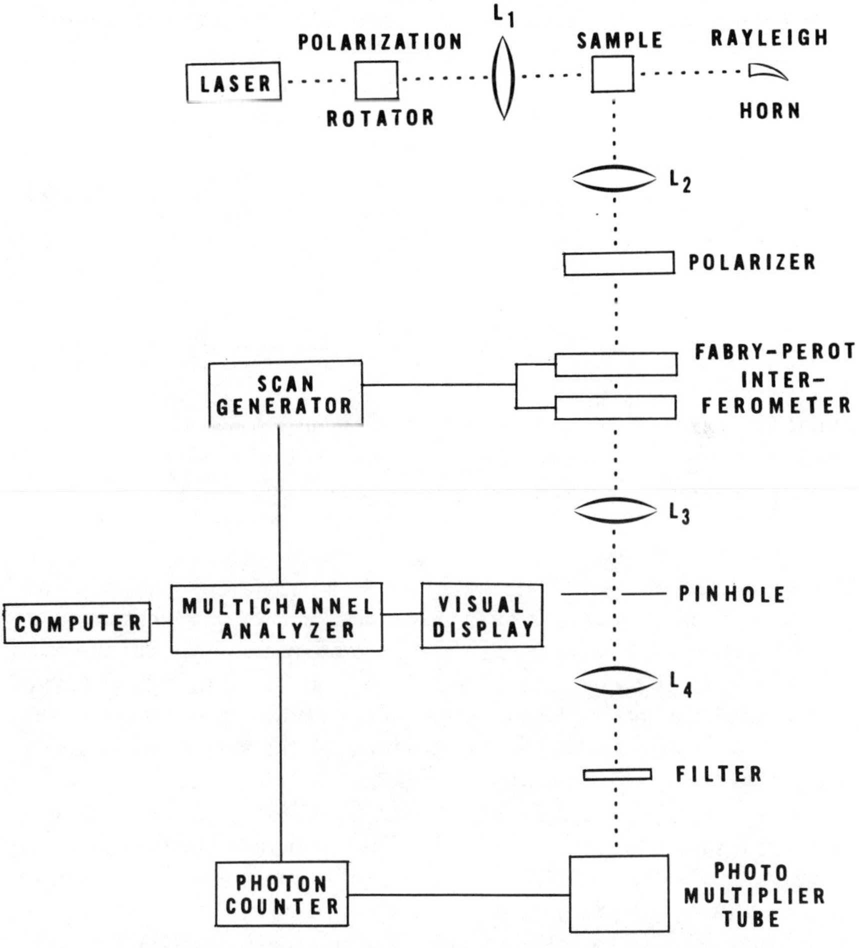

Figure 3. Block diagram of a light-scattering spectrometer

All the components necessary to assemble a complete Fabry–Perot spectrometer are available commercially. The total cost would be below $50,000. Thus, Brillouin scattering should become a more widely used technique in polymer science.

Examples and Discussion

The longitudinal modulus can be represented as:

$$M(\omega,p,T) = M_o(p,T) + \sum_j \frac{M_j(p,T)\,i\omega\tau_j(p,T)}{1 + i\omega\tau_j(p,T)} \tag{15}$$

where $M_o = K_o = \beta_T^{-1}$ is the thermodynamic value of the modulus given by the reciprocal of the isothermal compressibility, $\{M_j\}$ are the relaxing parts of the modulus, and $\{\tau_j\}$ are the relaxation times for the various modes of relaxation. If we consider only one relaxation time for K and one for G, and if we require that $\tau << 10^{-10}$ sec, then:

$$\pm\Delta\omega_{(1)} = q\left(\frac{\gamma}{\rho\beta_T}\right)^{1/2} \tag{16}$$

and

$$\Gamma_{(1)} = \frac{2q^2G_\infty\tau_S}{3\rho} + \frac{q^2K_R\tau_V}{2\rho} + \frac{\kappa q^2(\gamma - 1)}{2\rho C_P}$$

where G_∞ is the high-frequency shear modulus and $K_R = K_\infty - K_o$ is the relaxing part of the modulus of compression. We have taken $G_o = 0$ for a viscoelastic fluid. The shear viscosity is usually defined to be $\eta_S = G_\infty\tau_S$ and the volume viscosity $\eta_V = K_R\tau_V$. For most organic liquids, the part of Γ_1 caused by thermal conductivity is very small and will be neglected in the analysis.

If the density ρ and ratio of specific heats γ are known, measurements of $\Delta\omega_{(1)}$ can be used to obtain β_T. The ratio of specific heats for the case where $\tau << 10^{-10}$ sec can be obtained from the Rayleigh–Brillouin spectrum of the fluid. The intensity of the central peak owing to the thermal expansion divided by the intensity of the two Brillouin peaks is equal to $\gamma - 1$ (*6*). For *n*-hexadecane at 120°C (shown in Figure 1), this ratio yields $\gamma = 1.227$. The density is 0.7036 (*7*). The Brillouin splitting is measured to be 0.131 cm^{-1}. The isothermal compressibility is calculated to be 1.6×10^{-10} cm^2/dyn in good agreement with the directly measured value of β_T (*7*).

The longitudinal phonon velocity $V_{(1)}$ has been determined for the lower *n*-alkanes as a function of temperature by Champion and Jackson

(*8*). The higher *n*-alkanes and polyethylene (PE) have been studied by Patterson and Latham (*9*). There is a general relation between $V_{(1)}$ and relative free volume for all the homologs:

$$V_{(1)} \alpha \left(\frac{V_f}{V_o}\right)^{-1/3} \quad (17)$$

where $V_f = V - V_o$ is the Doolittle free volume (*10*). The results of Patterson and Latham are shown in Figure 4. The thermodynamic value of the phonon velocity $V_{(1)}(O) = (\gamma/\rho\beta_T)^{1/2}$ is insensitive to the detailed local structure of the homologous series of polymethylenes.

The Brillouin linewidth $\Gamma_{(1)}$ depends on the dynamic shear and volume viscosities $\eta_S(\omega)$ and $\eta_V(\omega)$. If the hypersonic shear viscosity $\eta_S(\Delta\omega_{(1)})$ is equal to the zero frequency shear viscosity and the small term caused by thermal conductivity is neglected, then measurements of $\Gamma_{(1)}$ can be used to obtain the volume viscosity. Champion and Jackson (*8*) noticed that the volume viscosities determined in the above manner for the *n*-alkanes were essentially independent of temperature. The values of $\Gamma_{(1)}$ measured in the author's laboratory for *n*-hexadecane are plotted

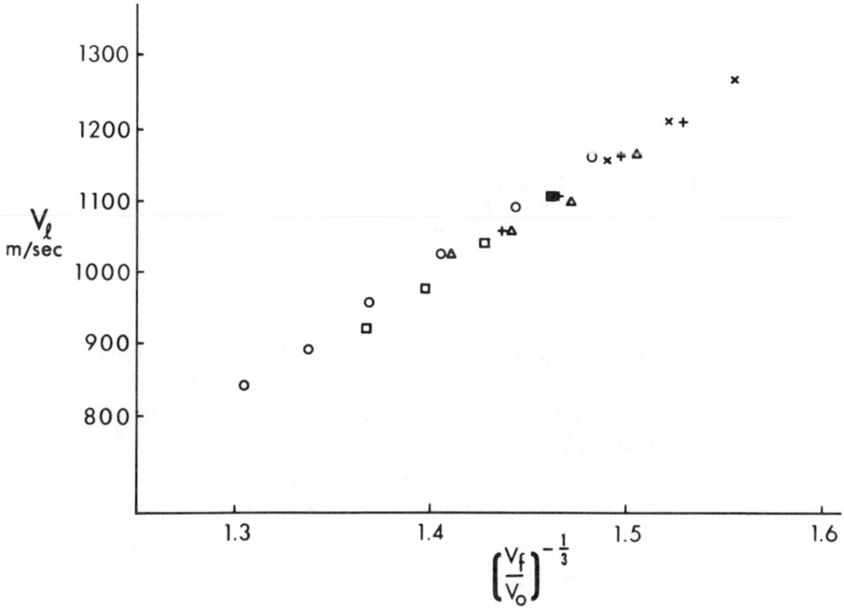

Figure 4. Longitudinal phonon velocity $V_{(1)}$ for several n*-alkanes and PE's vs.* $(V_f/V_o)^{-1/3}$*, where V_f is the free volume ($V_f = V - V_o$); ○ = C_{16}; □ = C_{36}; △ = P − 1000; + = P − 2000; × = PE.*

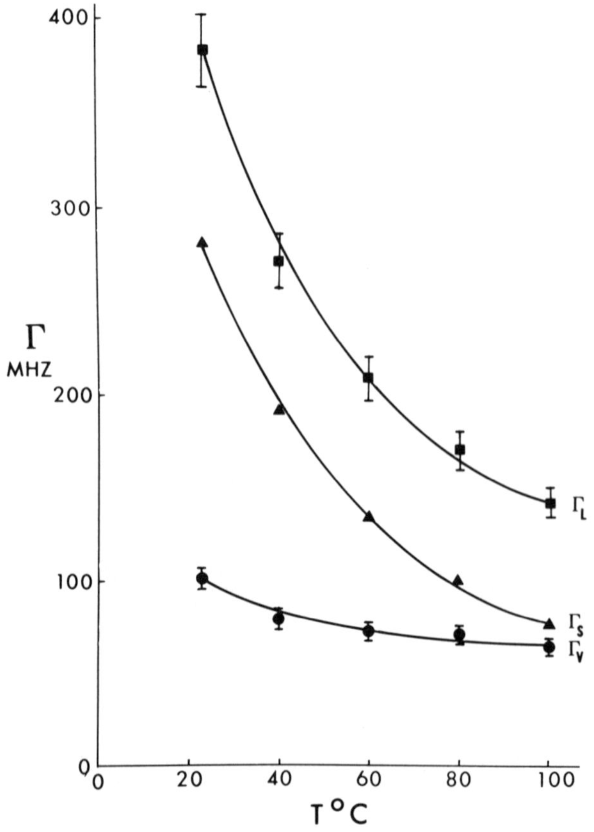

Figure 5. Longitudinal Brillouin linewidth $\Gamma_{(l)}$ for n-hexadecane vs temperature. Also plotted are the part of the linewidth owing to shear viscosity $\Gamma_S = [2q^2\eta_S(o)]/3\rho$ and the part caused by volume viscosity calculated as $\Gamma_V = \Gamma_{(l)} - \Gamma_S$.

versus temperature in Figure 5. In addition, the contributions to $\Gamma_{(1)}$ owing to shear (Γ_S) and volume (Γ_V) viscosity are presented. These results all assume that $\eta_S(\Delta\omega_{(1)}) = \eta_S(o)$ and they are in good agreement with Champion and Jackson (8).

However, there is compelling evidence that $\eta_S(\Delta\omega_{(1)}) < \eta_S(o)$ for *n*-hexadecane. The shear viscosity will have at least three contributions: (1) a translational structural relaxation, (2) overall molecular reorientation, and (3) intramolecular conformational relaxation. Thus $\eta_S(\omega)$ can be represented as:

$$\eta_S(\omega) = \frac{\eta_{trans}}{1 + \omega^2\tau^2_{trans}} + \frac{\eta_{rot}}{1 + \omega^2\tau^2_{rot}} + \frac{\eta_{conf}}{1 + \omega^2\tau^2_{conf}} \quad (18)$$

where τ_{trans}, τ_{rot}, and τ_{conf} are the relaxation times for the three contributions to the shear viscosity. Typical values of τ_{conf} for alkanes should be near 10^{-9} sec at room temperature (*11*). Thus, for Brillouin frequencies of 3×10^{10} rad/sec, the contribution of conformational relaxation to $\eta_S(\Delta\omega_{(1)})$ should be negligible.

Molecular reorientation times also have been determined for *n*-hexadecane (*21*). They range from 5×10^{-9} to 10^{-10} sec over the temperature interval of interest. Thus the contribution of molecular reorientation to $\eta_S(\Delta\omega_{(1)})$ also will be reduced greatly.

In order to obtain $\eta_S(\Delta\omega_{(1)})$ from $\eta_S(o)$, it is necessary to assess what fraction of $\eta_S(o)$ is caused by conformational relaxation and reorientational relaxation. This quantity has been obtained recently (*21*) and is near one-third. If we take into account the reduction in $\eta_S(\omega)$ at the Brillouin frequency $\Delta\omega_{(1)}$, the calculated hypersonic volume viscosity $\eta_V(\Delta\omega_{(1)})$ now depends on temperature and is comparable with $\eta_S(\Delta\omega_{(1)})$. The above result illustrates that the analysis of $\Gamma_{(1)}$ can be very complicated for chain-molecule fluids.

When $\tau_S << 10^{-10}$ sec, the transverse phonon velocity is imaginary and no transverse Brillouin peaks are observed. However, shear fluctuations do occur and they can couple with other modes of motion such as molecular reorientation. The spectrum that results is given by:

$$I_{HV}(\omega) \; \alpha \; \frac{\Gamma_{HV}}{\Gamma_{HV}^2 + \omega^2} \sin^2 \theta/2 + \Gamma_{HV} \cos^2 \theta/2 \, \frac{\frac{q^4\eta_S^2}{2}(1-R) + \omega^2}{\left(\Gamma_{HV}\frac{q^2\eta_S}{\rho} - \omega^2\right)^2 + \omega^2\left(\Gamma_{HV} + (1-R)\frac{q^2\eta_S}{\rho}\right)^2} \tag{19}$$

where $\Gamma_{HV} = (2\pi\tau_{or})^{-1}$ is the linewidth owing to overall molecular reorientation and R is a parameter which is equal to the fraction of the total shear viscosity which is caused by coupling to molecular reorientation. The depolarized Rayleigh–Brillouin spectrum of *n*-hexadecane at 65°C is shown in Figure 6 (*21*). The coupling leads to a central dip whose width is given by $q^2\eta_S/\rho$. A computer fit of the spectrum to the form of Equation 19 yields $\Gamma_{HV} = 1.30 \pm .05$ GHz, $q^2\eta_S/\rho = 196 \pm 20$ MHz, and $R = 0.33 \pm .02$. The calculated value of η_S based on the spectrum is in good agreement with the directly measured shear viscosity. The relaxation time for molecular reorientation is $\tau_{or} = .122 \times 10^{-9}$ sec.

As a liquid is cooled, the values of M_o, $\{M_j\}$, and $\{\tau_j\}$ increase. If the longitudinal Brillouin linewidth were given by Equation 16, $\Gamma_{(1)}$ would broaden and eventually exceed $\Delta\omega_{(1)}$. If we consider only a single struc-

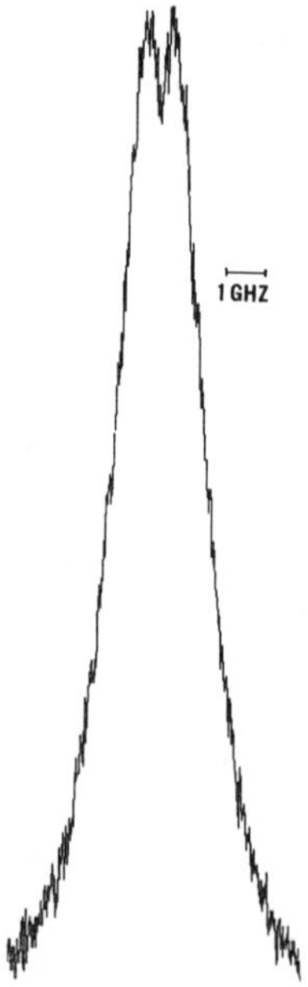

Figure 6. Depolarized (I_{HV}) Rayleigh–Brillouin spectrum of n-*hexadecane at 65°C*

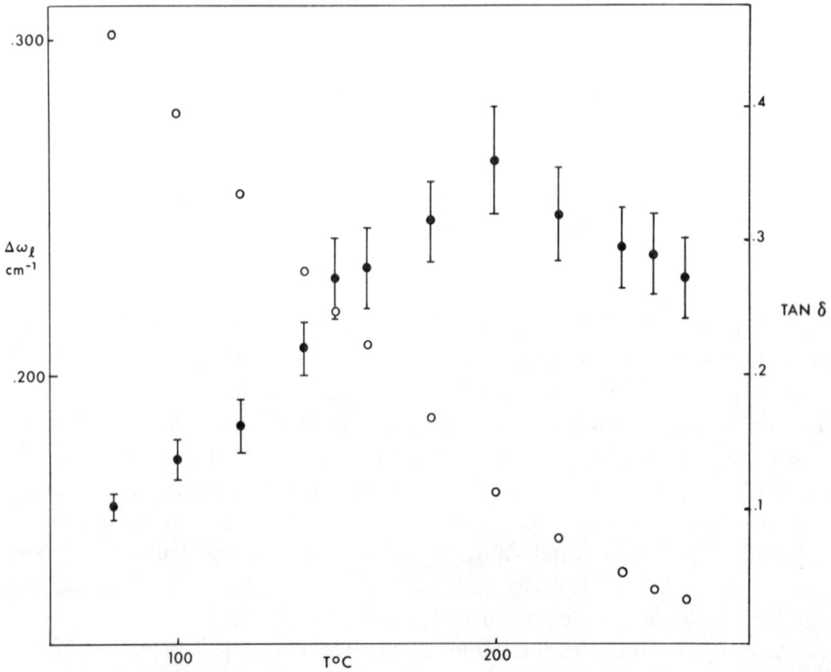

Figure 7. Brillouin splitting $\Delta\omega_{(l)}$(○) and tan δ(●) vs. temperature for polyisobutylene (PIB)

tural relaxation time and ignore contributions to $\Gamma_{(1)}$ owing to thermal conductivity or specific heat relaxation, then:

$$\Delta\omega_{(1)} \approx q\left(\frac{\gamma M_o}{\rho} + \left(\frac{\gamma K_R}{\rho} + \frac{4G_\infty}{3\rho}\right)\left(\frac{\Delta\omega_{(1)}{}^2\tau^2}{1+\Delta\omega_{(1)}{}^2\tau^2}\right)\right)^{1/2}$$

$$\Gamma_{(1)} \approx \frac{q^2\left(K_R + \frac{4}{3}G_\infty\right)^{\tau}}{2\rho(1+\Delta\omega_{(1)}{}^2\mathrm{T}^2)} \quad (20)$$

The linewidth will reach a maximum when $\Delta\omega_{(1)}\tau = 1$.

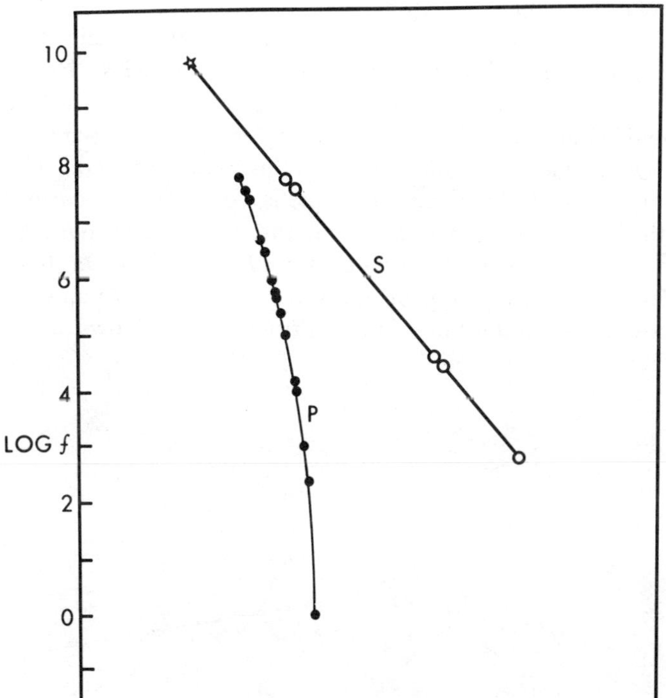

Figure 8. Transition map for PIB. The hypersonic result considerably extends the available frequency range. The primary (P) glass–rubber relaxation line and secondary (S) relaxation line are from Ref. 22.

Measurements of $\Delta\omega_{(1)}$ and $\Gamma_{(1)}$ in the relaxation region now have been carried out on a large number of polymers and temperatures of maximum loss have been determined. Values of $\Delta\omega_{(1)}$ and tan δ determined for polyisobutylene (PIB) (*12*) are plotted vs. temperature in Figure 7. The maximum loss is observed at approximately 200°C at a frequency of 4.95 GHz. Dynamic mechanical relaxation data often is presented on a so-called "transition map" where log f is plotted vs. $1/T_{max}$. The data usually group into discrete "transition lines." For amorphous polymers there is one line which correlates with the primary glass–rubber relaxation (P) and at least one secondary relaxation line (S) which tends to merge with the primary glass–rubber relaxation. The hypersonic results are combined with lower frequency data in Figure 8. The Brillouin results correlate very well with an extension of the secondary (S) glass–rubber relaxation line.

Usually the primary (P) glass–rubber relaxation cannot be resolved from the secondary relaxation at hypersonic frequencies. However, this is not always the case. The Brillouin frequencies $\Delta\omega_{(1)}$ and tan δ for polypropylene glycol (PPG) (*13*) are plotted versus temperature in Figure 9. Two tempratures of maximum loss are observed. The higher temperature loss at 100°C and a frequency of 4.40 GHz correlates very well with the primary glass–rubber relaxation line determined by dielectric relaxation at gigahertz frequencies (*13*). The lower temperature loss at 50°C and a frequency of 5.43 GHz correlates with an extension of the secondary transition line. The transition map is shown in Figure 10.

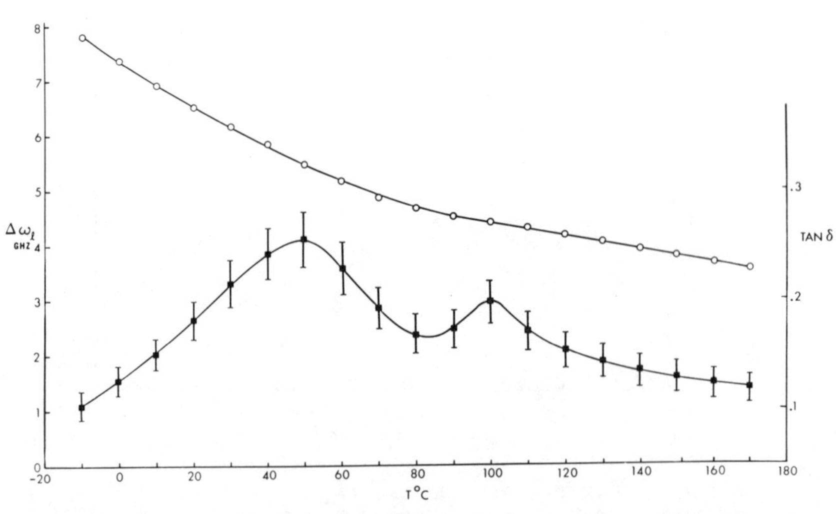

Figure 9. Brillouin splittings $\Delta\omega_{(l)}$(○) and tan δ(●) vs. temperature for poly-(d,(l) propylene glycol)

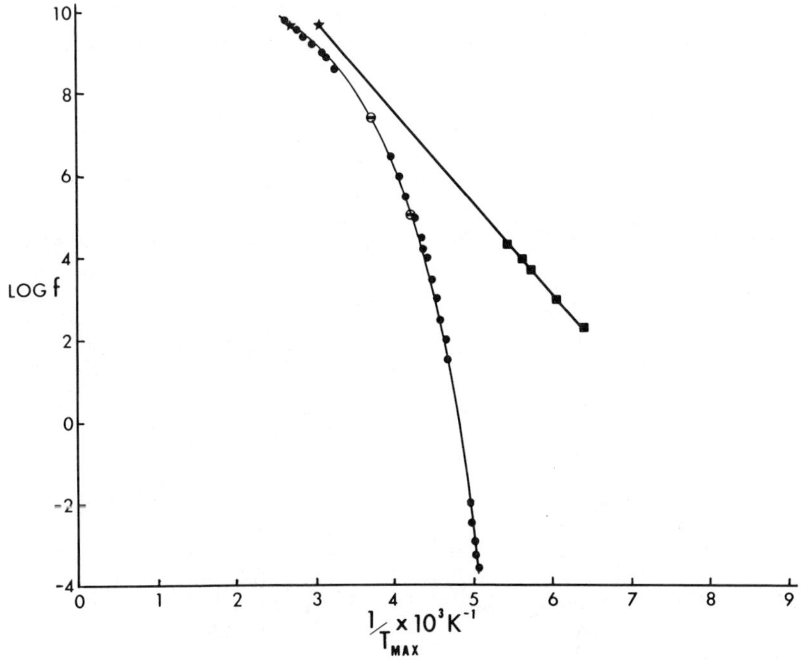

Figure 10. Transition map for PPG. The Brillouin results are indicated by ★.

It has been asserted (*14*) that above their melting points, the structural relaxation times in polymer fluids would be much less than 10^{-10} sec. This proved to be true for molten PE. However, for polyethylene oxide (PEO) a temperature of maximum loss was observed at approximately 60°C at a frequency of 6.06 GHz (*15*). The melting point of PEO is near 60°C. The temperature of maximum loss for bisphenol-A polycarbonate (*16*) was 280°C at a frequency of 5.43 GHz. The melting point of bisphenol-A polycarbonate is 240°C. Thus any general correlation between T_m and structural relaxation in fluids seems unwarranted.

In the relaxation region the transverse phonon velocity gives rise to a splitting that is comparable with the linewidth. The relations are:

$$\pm\Delta\omega_t = q\left(\frac{G_\infty}{\rho} - \frac{\left(\frac{1}{2\tau}\right)^2}{q^2}\right)^{1/2} \qquad (21)$$

$$\Gamma_t = \frac{1}{2\tau}$$

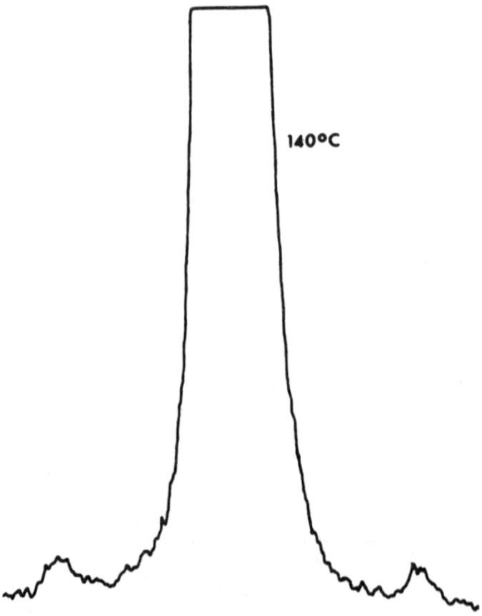
140°C

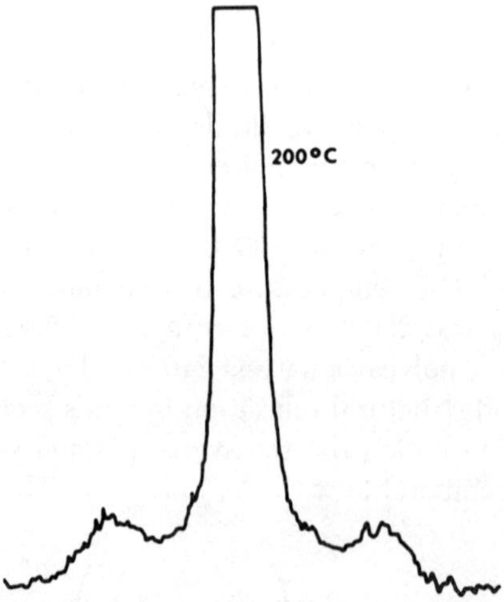
200°C

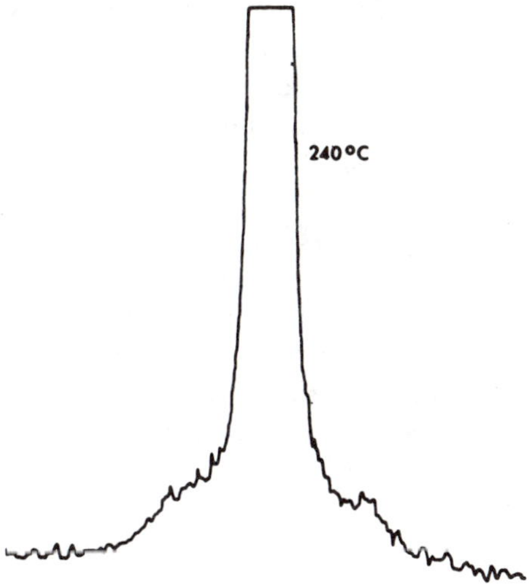

Figure 11. Depolarized (I_{HV}) Rayleigh–Brillouin spectrum of bisphenol-A polycarbonate at 140° (upper left), *200°* (lower left), *and 240°C* (top)

where $\Delta\omega_t$ and Γ_t are expressed in rad/sec. If $\tau = 5 \times 10^{-11}$ sec and $G_\infty/\rho = 5 \times 10^9$ cm²/sec², then $\Delta\omega_t = 2.4$ GHz and $\Gamma_t = 1.6$ GHz. The depolarized (I_{HV}) spectrum of bisphenol-A polycarbonate at several temperatures is shown in Figure 11. At $T_{(g)}$ (140°C), the transverse peaks are well defined. At 200°C the peaks are much broader. At 240°C the width and splitting are becoming comparable. There is also a large central peak in the depolarized spectrum.

When the structural relaxation time $\tau >> 10^{-10}$, the Brillouin splittings reach their limiting values:

$$\pm\Delta\omega = qV_\infty \tag{22}$$

where V_∞ is the high-frequency, limiting-phonon velocity. The value of V_∞ for an equilibrium fluid depends only on pressure and temperature. As the relaxation time approaches 10^3 sec, the fluid will not reach the true equilibrium state if it is heated or cooled at 10°C/min, a rate typical for a differential scanning calorimeter. The fluid will exhibit macroscopic hysteresis and the low-temperature material will be called a glass. However, if the fluid is allowed to attain equilibrium at each temperature the observed phonon velocity should fall on an extension of the equilibrium liquidous line. Measurements of $\Delta\omega_{(1)}$ for atactic poly(methyl methacrylate) (PMMA) and for two atactic polystyrenes (PS's) in the region of the glass transition are shown in Figure 12 (5). The value of $T_{(g)}$, determined with a differential scanning calorimeter, is indicated by an arrow. The samples were allowed to reach equilibrium at each temperature and thus there is no change in slope of $\Delta\omega_{(1)}$ versus temperature at the nominal $T_{(g)}$.

Many crystallizable polymers can be prepared in the amorphous glassy state by rapid quenching as films. Measurements of $\Delta\omega_{(1)}$ or $\Delta\omega_t$ can then be made as a function of temperature and the glass transition temperature determined. Such results are shown for amorphous polyethylene terephthalate (PET) in Figure 13 (*17*). The Brillouin splittings change slope at 70°C. If both $\Delta\omega_{(1)}$ and $\Delta\omega_t$ can be measured, the Poisson ratio σ can be determined according to:

$$\sigma = \frac{\frac{1}{2}\left(\frac{\Delta\omega_{(1)}}{\Delta\omega_t}\right)^2 - 1}{\left(\frac{\Delta\omega_{(1)}}{\Delta\omega_t}\right)^2 - 1} \tag{23}$$

For many samples, Brillouin scattering may be the easiest way to obtain σ.

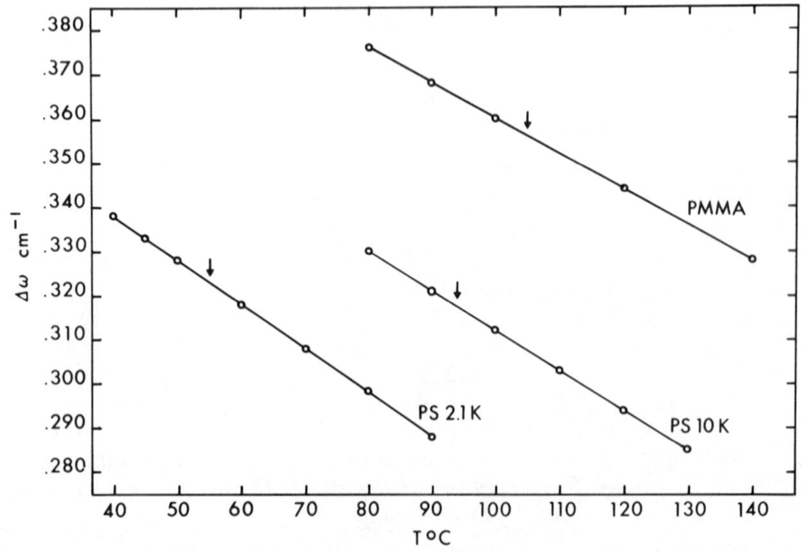

Figure 12. Brillouin splittings $\Delta\omega_{(1)}$ vs. temperature near the glass–rubber relaxation for PMMA, 10,000 molecular-weight PS and 2100 molecular-weight polystyrene. The arrows indicate the value of $T_{(g)}$ determined with a differential scanning calorimeter.

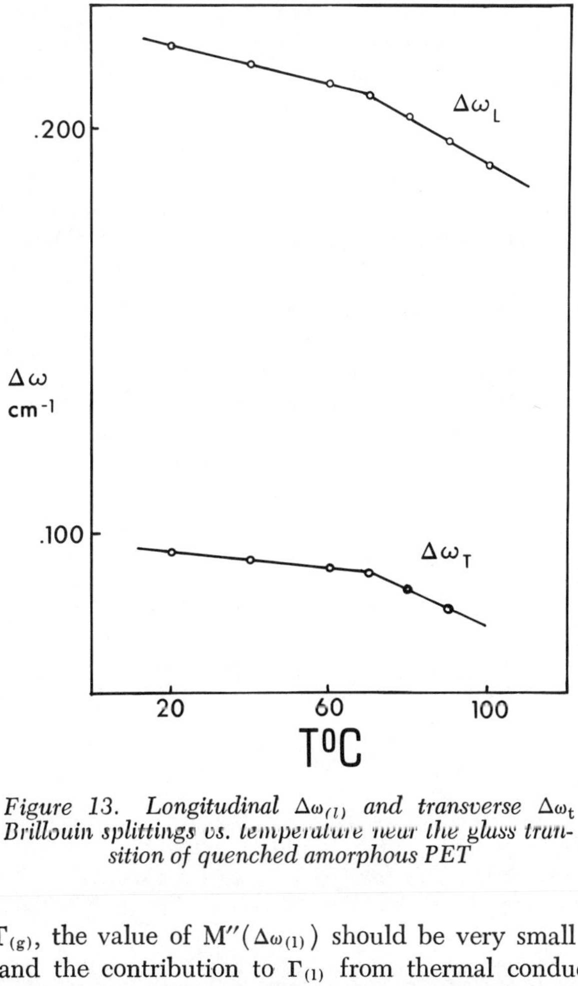

Figure 13. Longitudinal $\Delta\omega_{(l)}$ and transverse $\Delta\omega_t$ Brillouin splittings vs. temperature near the glass transition of quenched amorphous PET

Near $T_{(g)}$, the value of $M''(\Delta\omega_{(1)})$ should be very small relative to $M'(\Delta\omega_{(1)})$ and the contribution to $\Gamma_{(1)}$ from thermal conductivity also should be quite small. However, the observed linewidths in polymers near the glass transition are in the range 100–300 MHz (*18, 19*). The broad lines are caused by density fluctuations which persist for times long relative to the period of a longitudinal phonon.

The simple linear response theories presume that the fluid is at equilibrium when the phonon is created or destroyed. However, when $\tau >> 10^{-10}$ sec this assumption is not valid. The Brillouin spectrum actually observed is the sum of the spectra from many small volumes in the sample. If we assume that the intrinsic linewidth of the Brillouin peaks is negligible, the observed spectrum should be given by:

$$I_B(q,\Delta\omega) \propto \exp - \left(A \frac{((\Delta\omega)^2 - <\Delta\omega_{(1)}>^2)^2}{(<\rho^2> - <\rho>^2)} \right) \qquad (24)$$

where $<\Delta\omega_{(1)}>$ is the average Brillouin splitting corresponding to the equilibrium density. The width of the peaks is proportional to the mean-squared density fluctuations. Below the glass transition region, the non-propagating density fluctuations essentially are frozen in and the linewidth should change only slightly below $T_{(g)}$. This effect is illustrated for PMMA in Figure 14.

Polymer blends are seldom compatible, but when they are the longitudinal Brillouin peaks should be as narrow as those observed in pure amorphous materials. The Brillouin spectrum of a film of PMMA at $T_{(g)} = 100°C$, showing two Fabry–Perot orders is shown in Figure 15. The peaks are seen to be quite sharp. Films of PMMA and poly(vinylidene fluoride) (PVF_2) are (*20*) compatible above the melting point of PVF_2 and amorphous glassy films can be prepared by rapid quenching. The Brillouin spectrum of a quenched film containing 75% PVF_2 and

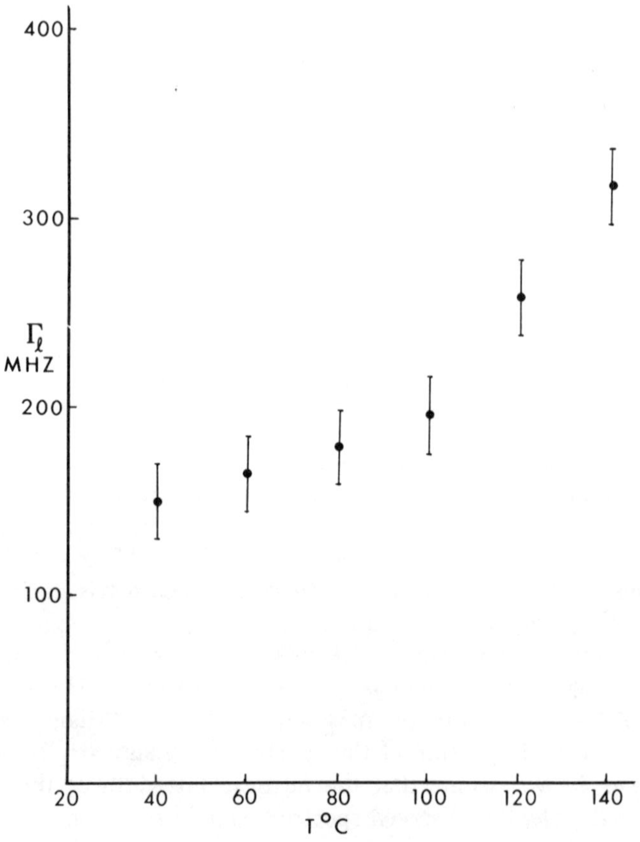

Figure 14. Longitudinal Brillouin linewidths $\Gamma_{(l)}$ vs. temperature for PMMA near $T_{(g)}$

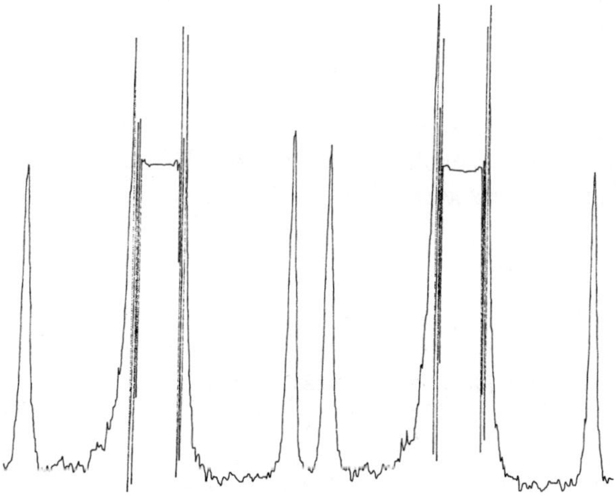

Figure 15. Brillouin spectrum of a film of PMMA at 100°C. Two Fabry–Perot orders are shown.

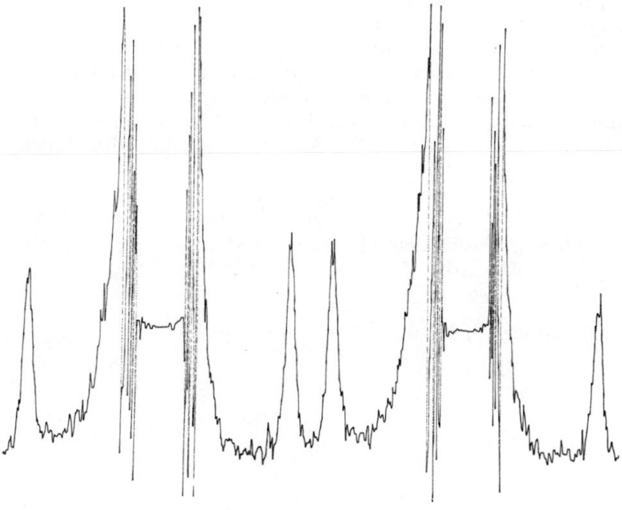

Figure 16. Brillouin spectrum of a quenched film of 75% PVF_2 and 25% PMMA at 20°C

25% PMMA by weight at its $T_{(g)} \approx 20°C$ is shown in Figure 16. The sharp peaks confirm that the sample is a homogeneous amorphous mixture. Measurements of the Brillouin linewidth are a sensitive measure of the homogeneity of the sample.

Brillouin scattering is a versatile technique for studying the physical properties of polymers. Both thermodynamic and kinetic information can be obtained. Even morphological studies can be carried out. The use of Brillouin scattering as a standard tool in polymer science now seems assured.

Literature Cited

1. Einstein, A., *Ann. Phys.* (1910) **33,** 1275.
2. Brillouin, L., *Ann. Phys. Paris* (1922) **17,** 88.
3. Leontovich, M. A., *Z. Phys.* (1931) **72,** 247.
4. Sandercock, J. R., *Opt. Commun.* (1970) **2,** 73.
5. Patterson, G. D., *J. Polym. Sci., Polym. Lett. Ed.* (1975) **13,** 415.
6. Landau, L., Placzek, G., *Phys. Z. Sowjetunion* (1934) **5,** 172.
7. Orwoll, R. A., Flory, P. J., *J. Am. Chem. Soc.* (1967) **89,** 6814.
8. Champion, J. V., Jackson, D. A., *Mol. Phys.* (1976) **31,** 1169.
9. Patterson, G. D., Latham, J. P., *Macromolecules* (1977) **10,** 736.
10. Doolittle, A. K., *J. Appl. Phys.* (1951) **22,** 1471.
11. Piercy, J. E., Rao, M. G. S., *J. Chem. Phys.* (1967) **46,** 395.
12. Patterson, G. D., *J. Polym. Sci., Polym. Phys. Ed.* (1977) **15,** 455.
13. Patterson, G. D., Douglass, D. C., Latham, J. P., *Macromolecules* (1978) **11,** 263.
14. Starkweather, H. W., Jr., *J. Macromol. Sci., Phys.* (1968) **B2,** 781.
15. Patterson, G. D., Latham, J. P., *Macromolecules* (1977) **10,** 1414.
16. Patterson, G. D., *J. Macromol. Sci., Phys.* (1977) **B13,** 647.
17. Patterson, G. D., *J. Polym. Sci., Polym. Phys. Ed.* (1976) **14,** 1909.
18. Huang, Y. Y., Friedman, E. A., Andrews, R. D., Hart, T. R., "Light Scattering in Solids," M. Balkanski, Ed., p. 488, Flammarion, Paris, 1971.
19. Jackson, D. A., Pentecost, H. T. A., Powles, J. G., *Mol. Phys.* (1972) **23,** 425.
20. Patterson, G. D., Nishi, T., Wang, T. T., *Macromolecules* (1976) **9,** 603.
21. Patterson, G. D., Alms, G. R., *Macromolecules* (1977) **10,** 1237.
22. McCall, D. W., "Molecular Dynamics and Structure of Solids," R. S. Carter, J. J. Rush, Eds., *Nat. Bur. Stand. U.S. Spec. Publ.* (1969) **301,** 475–537.

RECEIVED November 10, 1977.

10

Structural Characterization of Polymers in Solution by Quasielastic Laser Light Scattering

A. M. JAMIESON

Department of Macromolecular Science, Case Western Reserve University, Cleveland, OH 44106

M. E. McDONNELL

Department of Physics, West Virginia University, Morgantown, WV 26506

The theoretical and experimental basis for quasielastic laser light scattering (QLLS) spectroscopy is reviewed. A diverse range of information regarding the structural characteristics of polymer solutions is accessible through application of QLLS techniques. The translational diffusion coefficient of polymers can be determined speedily and accurately. From the limiting value D_t^0 *and from the concentration dependence* dD_t/dC *details of molecular weight, size and shape, flexibility, charge density, and porosity of the polymer can be obtained. Rotational diffusion coefficients of rigid anisometric particles and relaxation times for intramolecular motions of flexible polymers also can be measured. In more concentrated solutions where intermolecular interactions modify the free-particle diffusion behavior, the QLLS technique helps to gain information about the range and character of intermolecular forces or the presence of intermolecular entanglements of flexible coil macromolecules.*

Interest in the investigation of macromolecular diffusion phenomena in both dilute and concentrated solutions has received recent stimulus with the development of techniques which can conveniently and accurately identify the dynamic characteristics of light scattered from fluctuations in solute concentration. In dilute solution, part of this activity stems from the fact that the limiting value of the translational diffusion

0-8412-0406-3/79/33-174-163$10.00/1

coefficient of macromolecular species D_t° can be determined quickly and accurately by this new technology. Thus D_t° can be conveniently used to probe the structural features of the macromolecular solute, either by itself or combined with independent thermodynamic or hydrodynamic measurements. On the other hand, studies of the dependence of the diffusion coefficient on concentration and scattering angle enable characterization of the range and dynamics of intermacromolecular interactions. In contrast to those discussed here, classical techniques to measure diffusion coefficient are limited by low accuracy and/or time-consuming experimentation. The quasielastic laser scattering (QLLS) techniques discussed here are also denoted generically by the term dynamic light scattering (*1*) and include the individual techniques referred to as optical mixing, optical homodyne or heterodyne, light beating, photon correlation, and intensity fluctuation spectroscopy. These methods have found widespread application in areas of materials science, biology, and physical chemistry of fluids and interfaces. These developments have been extensively reviewed elsewhere (*2–12*). This chapter critically explores their usefulness for characterizing polymer structure in solutions. The structural parameters in question are those which relate to macromolecular dimensions, conformational forms, degree of association, or intermolecular ordering in solution. This review provides a short description of the theoretical basis for the quasielastic light scattering method and the instrumentation currently available for observing the dynamic behavior of scattered laser light. Recent applications of the techniques to the structural analysis of polymer solutions are reviewed. The parameters discussed in this chapter pertain to the hydrodynamic structure of the macromolecular solutions since the information obtained by these investigations characterize the motion of the macromolecules in a solvent. Dimensional information about the polymer will therefore be affected by properties such as porosity, deformability (related to the hydrodynamic interactions of chain segments), and degree of solvation. We will also survey the information about intermolecular hydrodynamic structure in concentrated solutions that can be gleaned by the new light scattering methods. These investigations concern information about the correlation lengths for intermolecular interactions, the nature of the intermolecular force field, and the presence of entanglements of flexible coil polymers.

Theoretical Considerations

The theoretical basis of the quasielastic light scattering phenomenon has been explored thoroughly in reviews (*2–12*) as well as in three volumes (*1, 13, 14*), and we will summarize the salient features in this

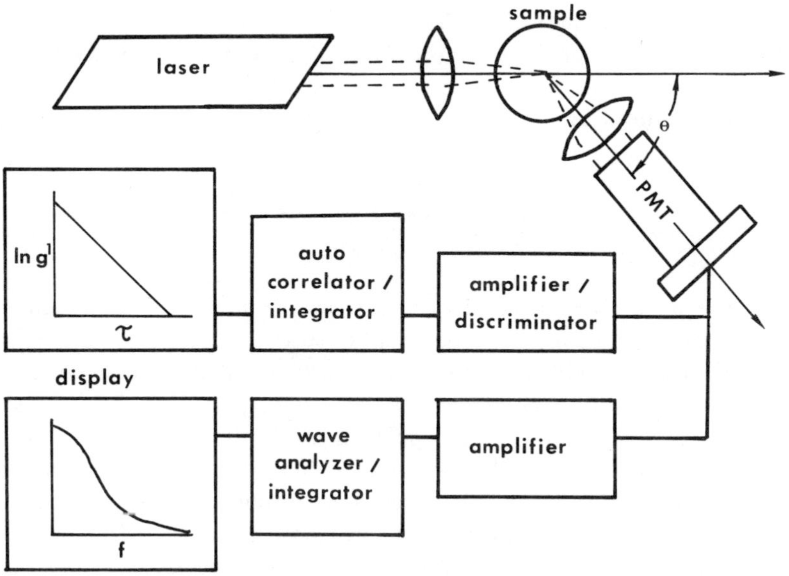

Figure 1. Schematic of instrumentation required for quasielastic laser light scattering spectroscopy

section. The scattering of light from any material occurs because of microscopic, thermally driven fluctuations in the refractive index $\delta\bar{n}(r,t)$. In polymer solutions, the overwhelming contribution to $\delta\bar{n}(r,t)$ arises from the Brownian fluctuations in concentration of the components $\delta c(r,t)$. For the usual experimental configuration encountered in laser light scattering experiments and schematized in Figure 1, with the polarization of incident light perpendicular to the plane defined by the direction of propagation of incident light and the direction of scattered beam, the intensity of scattered light is proportional to the mean square refractive index fluctuation. The excess light scattering intensity of the solution over the solvent is therefore proportional to the mean square fluctuation in solute concentration. This quantity is determined by the concentration derivative of the chemical potential which, in turn, is related to the osmotic compressibility at constant pressure, temperature, and constant chemical potential of solvent species $(\partial\pi/\partial c_2)_\mu$ where c_2 is the concentration of polymer in g/cc. Thus:

$$\frac{K_\mu c_2}{\Delta R} = \left(\frac{\partial\pi}{\partial c_2}\right)_\mu \tag{1}$$

$$= M_2^{-1} + 2A_{2\mu}c_2 + \ldots \tag{2}$$

where the Rayleigh factor $R = r^2 i_s / I_0$, r is the distance from the scattering center to the detector, i_s/I_0 is the ratio of scattered to incident light intensity, Δ signifies that the increment of R between solution and solvent is measured, $A_{2\mu}$ is the second osmotic virial coefficient at constant chemical potential, and K_μ is an optical constant:

$$K_\mu = \frac{2\pi^2 \bar{n}^2 (\partial \bar{n}/\partial c_2)_\mu{}^2}{N_A \lambda^4} \quad (3)$$

where $\bar{n}$ is refractive index, $(\partial \bar{n}/\partial c_2)_\mu$ is the refractive index increment measured at constant chemical potential of solvent species, λ is the wavelength of incident light, and N_A is Avogadros number. When polymers are dissolved in multicomponent solvents (e.g., polyelectrolytes in salt solutions), as the concentration of polymer is varied, the chemical potential of solvent species may change if they exhibit preferential interactions with the polymer (*15, 16*). This leads to erroneous extrapolations to zero concentration unless appropriate corrections are made (*15, 16*). The simplest procedure is to exhaustively dialyze the solutions against the solvent system at each polymer concentration and to measure ΔR and $(d\bar{n}/dc_2)_\mu$ between the dialyzed solution and dialysate (*15, 16*). Equation 1 is the basis of the application of the excess intensity light scattering method to molecular weight determination, but it applies only when the macromolecular size is small compared with λ. Otherwise the scattered intensity becomes angle dependent because of interference at the detector between light scattered from different points on the same molecule. This variation is described by the intramolecular particle scattering function $P(\theta)$:

$$\frac{K_\mu c_2}{\Delta R} = \frac{1}{P(\theta)}\left[\frac{1}{M_2} + 2A_{2\mu}c_2 + \ldots\right] \quad (4)$$

and M_2 must be evaluated in the limit $\theta \rightarrow 0$ where $P(0) = 1$. Additional corrections must be made if the macromolecular solute is optically anisotropic and scatters significant amounts of depolarized light since Equation 4 refers only to the isotropic component of the scattered light (*17*).

Quasielastic laser light scattering technology permits an analysis of the dynamic behavior of the scattered light which is determined by the dynamics of the concentration fluctuations. Experimentally, the quantities which may be determined are, therefore:

(1) The intensity i_s, defined as the mean square of the scattered electric field:

$$i_s = \langle |E_s|^2 \rangle \quad (5)$$

where

$$E_s = \sum_j A_j(t) e^{i\vec{q}\cdot\vec{r}_j(t)} e^{-i\omega_0 t} \quad (6)$$

The scattering amplitude $A_j(t)$ of the j'th scattering element will change with time only if the scattering element experiences an internal relaxation (structural, or orientational if the particles are anisotropic); r_j is the position of the j'th scattering element; q is the scattering vector defined as the change in the wave vector k between the incident and scattered light:

$$|q| = |k_o - k_s| \approx 2|k_o|\sin\theta/2$$
$$= (4\pi\overline{n}/\lambda)\sin\theta/2 \quad (7)$$

where the scattering angle θ is defined in Figure 1; the phase factor $e^{i\vec{q}\cdot\vec{r}(t)}$ contains information about the translational motions of the scattering centers; and ω_o is the frequency of incident radiation.

(2) The autocorrelation function of the scattered field:

$$G^{(1)}(\tau) = \langle E_s^*(t) E_s(t+\tau) \rangle \quad (8)$$

which describes the dynamic behavior imposed on $E_s(t)$ by the dynamics of the scattering centers, i.e., it characterizes the relation of the magnitude of E_s at time t to the magnitude at a later time $t + \tau$;

(3) The optical spectrum, the Fourier transform of $G^{(1)}(\tau)$:

$$I(\omega) = \frac{1}{2\pi}\int_{-\infty}^{\infty} G^{(1)}(\tau)\, e^{i\omega\tau} d\tau \quad (9);$$

(4) The intensity autocorrelation function:

$$G^{(2)}(\tau) = \langle i^*(t) i(t+\tau) \rangle \quad (10);$$

(5) The power spectrum $P_i(\omega)$ which is the Fourier transform of $G^{(2)}(\tau)$:

$$P_i(\omega) = \frac{1}{2\pi}\int_{-\infty}^{\infty} G^{(2)}(\tau)\, e^{i\omega\tau} d\tau \quad (11)$$

In homodyne spectroscopy one measures either $G^{(2)}(\tau)$ or $P_i(\omega)$; in heterodyne spectroscopy $G^{(1)}(\tau)$ or $I(\omega)$ is obtained. In the homodyne experiment only the light scattered by the system under investigation is permitted to reach the phototube; in heterodyne measurements the light scattered by the system is permitted to mix coherently with a portion of unbroadened laser light before it reaches the detector. In all of the situations encountered in this review, the scattered field represents a stationary Gaussian random process, and information gained by the homodyne and heterodyne techniques is entirely equivalent (*1*). That is to say:

$$g^{(2)}(\tau) = 1 + |g^{(1)}(\tau)|^2 \tag{12}$$

where

$$g^{(1)}(\tau) = G^{(1)}(\tau)/G^{(1)}(0)$$

and

$$g^{(2)}(\tau) = G^{(2)}(\tau)/G^{(1)}(0)^2$$

are the normalized forms of the autocorrelation functions. The simplest theoretical case corresponds to that in which the scattering volume contains N identical spherical scatterers, each with a time-independent scattering amplitude A, and the field autocorrelation function can be written:

$$G^{(1)}(\tau) = < \sum_{j=1}^{N} Ae^{-i\vec{q}\cdot\vec{r}_j(0)} \sum_{j'=1}^{N} Ae^{i\vec{q}\cdot\vec{r}_{j'}(\tau)} > e^{-i\omega_0\tau} \tag{13}$$

For systems which are sufficiently dilute so that the positions r_j and r_j' of different particles are uncorrelated but not so dilute that the scattered field may become non-Gaussian, the cross-terms in jj' have zero averages, and

$$G^{(1)}(\tau) = N|A|^2 < e^{iq\cdot\vec{r}(\tau)-\vec{r}(0)} > e^{-i\omega_0\tau} \tag{14}$$

$$= N|A|^2 \int G_s(\vec{R}, \tau) e^{i\vec{q}\cdot\vec{R}} dR^3 \cdot e^{-i\omega_0\tau} \tag{15}$$

where $G_s(\vec{R}, \tau)$ is the conditional probability that a particle located at the origin at time zero will be at the position $\vec{R}$ at time τ, and represents the self part ofthe VanHove space–time correlation function (*18*). When finite correlation exists between individual scatterers j and j', the discrete contribution to the space–time correlation function becomes nonzero, as we will discuss later, and more complex dynamic behavior of the scattered light is encountered.

For particles involved in free isotropic translational diffusion, G_s is described by Fick's equation for translational diffusion:

$$\frac{\partial G_s}{\partial t} = D_t \nabla^2 G_s \tag{16}$$

which on integration and insertion in (14) and (15) leads to:

$$G^{(1)}(\tau) = N|A|^2 e^{-D_t q^2 \tau} e^{-i\omega_0 \tau} \tag{17}$$

From Equation 9, the optical spectrum is thus:

$$I(\omega) = N|A|^2 \left[\frac{D_t q^2/\pi}{(\omega - \omega_0)^2 + (D_t q^2)^2} \right] \tag{18}$$

which corresponds to a Lorentzian function centered on ω_0 with half-width at half-height

$$\Delta\omega_{\frac{1}{2}} = D_t q^2 \tag{19}$$

The corresponding relations for the normalized second-order correlation function can be shown to be:

$$g^{(2)}(\tau) = 1 + e^{-2D_t q^2 \tau} \tag{20}$$

and, for the power spectrum:

$$P_i(\omega) = N^2|A|^4 \left[\frac{2D_t q^2/\pi}{(\omega)^2 + |2D_t q^2|^2} \right] \tag{21}$$

If the scatterers are nonspherical and optically isotropic, the scattering amplitudes A_i become time dependent. If the position and orientation of single and pairs of scatterers are uncorrelated, the average of the time-dependent scattering amplitude can be separated from the position average:

$$G^{(1)}(\tau) = N < A(0) A(\tau) > \quad < e^{iq \cdot [r(\tau) - r(0)]} > e^{-i\omega_0 \tau}$$

where $A(\tau) = A(0) + A_1(\tau)$. It can be shown that the effect of this is to mix in contributions of orientational and translational diffusion (*19*) so that for cylindrically symmetric particles of length L:

$$G^{(1)}(\tau) = N A_{2q=0} [B_0 e^{-D_t q^2 \tau} + B_2 e^{-(D_t q^2 + 6D_r)\tau} + \ldots] e^{-i\omega_0 \tau} \tag{22}$$

where D_r is the rotational diffusion coefficient. The rotational term in Equation 22 only becomes significant compared with the first term when $qL > 5$; at small angles only pure translational diffusion contributes to the scattering; at $qL = 10$ about 50% of the total scattering is composed of the first term in Equation 22 if the particles resemble long thin rods. Of course the total scattering is at the same time reduced at higher angles because of the internal interference effect which is characterized by $P(\theta)$. The approximation of uncorrelated orientation and position is somewhat of an oversimplification (*20*), but the important point is that at wide angles when the scatterers are large, rigid, and anisometric, there will no longer be single exponential character to $G^{(1)}(\tau)$ and $G^{(2)}(\tau)$ or single Lorentzian character to $I(\omega)$ or $P(\omega)$, at the same time, the total scattering intensity will be decreased.

If the scattering particles are optically anisotropic, there is, in addition to the isotropic scattering components described above, a depolarized component (*21*):

$$G^{H(1)}(\tau) = N < A^{H}(0) A^{H}(\tau) > e^{-D_t q^2} e^{-i\omega_0 \tau} \quad (23)$$

$$= N |A^{H}(0)|^2 e^{-(D_t q^2 + 6D_r)} e^{-i\omega_0 \tau} \quad (24)$$

i.e., there is no pure translational term in the depolarized scattering contribution. Also, the amplitude factor $|A^H(0)|^2$ is a function of the optical anisotropy only, not the molecular dimensions, and so is constant with scattering angle.

Instrumental Detection

As in most fields of investigation, there are many kinds of equipment available for use in dynamic light scattering depending on the intended use, the components at hand, the facilities for local fabrication, and the money available. The optimum configuration of a spectrometer depends on the system to be studied, but much useful information can be collected for many scattering systems from any one of several carefully conceived instruments (*13*). Each photometer can be divided conveniently into four systems: the laser light source, scattering chamber, detector, and signal processor. For most applications good components are commercially available. To describe the variety of QLLS photometers commonly used, it is easiest to discuss the system in the direction opposite from that which the light travels; that is, starting with the signal analyzer.

Originally the radiofrequency spectrum of the output voltage of the optical mixing photomultiplier tube was analyzed with a swept filter analyzer one frequency band at a time, to determine $I(\omega)$ or $P_i(\omega)$. With this equipment, satisfactory signal-to-noise ratio may take several hours, and so not only all the components but also the system being studied have to have good long-term stability. Alternatives became available with dedicated computer technology: real time signal analyzers and autocorrelators. The latter instruments measure $G^{(1)}(\tau)$ or $G^{(2)}(\tau)$. By analyzing the data collected for all frequencies as quickly as it is collected, good signal-to-noise ratios can be obtained in several minutes, and the stability requirements are relaxed. Real-time spectrum analyzers present the data in the way it had traditionally been viewed and permit simple visual removal of unwanted harmonic components of noise spectrum (e.g., line pick-up). The most efficient way to use a spectrum analyzer is to determine the voltage spectrum of the photocurrent which corresponds to the square root of $I(\omega)$ or $P_i(\omega)$, depending on whether a homodyne or heterodyne experiment is enacted.

Correlators have the advantage that they do not use wide-band amplifiers with their inherent problems of distortion. In addition, the correlator can be used in conjunction with a suitable photomultiplier tube, pulse amplifier, and discriminator combination (which provides the input signal as discrete amplified photon pulses) to determine the correlation function of the incident photon flux. Under these circumstances, the incoming pulses are counted and processed in a completely digital fashion, resulting in increased precision. For many applications, it is sufficient to determine the clipped autocorrelation function instead of the full correlation function (Ref. *13*, Chapter VI). The number of incoming photons in a finite time interval is assigned a value of 1 or 0 depending on whether it exceeds a particular value (clipping level). The clipped function can be computed more simply and much more quickly than the full correlation function, permitting analysis of decay rates several orders of magnitude faster than the latter. The requirement for no loss of information in the substitution of a clipped for a full correlation function is that the scattering process be Gaussian random. In such a case, the scattering system can be described fully by the clipped function; if the scattering system provides non-Gaussian signal statistics, clipped correlation functions, computed by summing over a variable (Ref. *13*, Chapter VI) or random clipping level (*22*) also provide a method to measure the full photocurrent correlation function. Non-Gaussian correlation functions arise at very low concentrations because the number of particles in the scattering volume have a Poisson distribution. In such a case, the homodyne intensity correlation function decays on two widely different time scales. The fast process is the usual Brownian diffusion motion of individual particles, the slow process is related to the time dependence of the total number of particles in the scattering volume (*23*). The slow relaxation does not appear in the heterodyne experiment. The number fluctuation phenomenon can provide additional information on the colloid statistics (*12*).

The detector for a QLLS spectrometer is always a high-gain photomultiplier tube. This tube produces an electric signal which is proportional to the light intensity falling on its photocathode. Typically only one photon in 10 will knock an electron out of the photocathode. The tube will then amplify this signal of the order 10^7 times. To produce a signal for analog analysis the voltage is sampled across a large load resistor. Fluctuations in the incident light intensity result in a voltage which fluctuates around a mean value that corresponds to the average intensity of light scattered. When photon counting equipment is used a small load resistor facilitates separation of the individual pulses coming from the tube. These pulses are further amplified and processed through a discriminator which excludes noise pulses resulting from electrons which do not originate from the photocathode. The output of the discriminator is a uniform pulse for each acceptable incoming pulse. By using pulses of uniform shape the uncertainties that would be introduced from processing signals with slightly different amplifications in the analog signal processing are eliminated.

Selecting the best photomultiplier tube can be frustrating since different manufacturers do not give a complete list of comparable specifications. If analog signals are being detected, a tube is selected with high gain and quantum efficiency and low dark current. This generally means selecting a high gain tube with a spectral response that has its maximum efficiency near the wavelength of interest and as sharp a cut off on the red response as possible. For photon counting, which involves signals with lower intensities, all these qualities are sought, but in addition it becomes important that the photomultiplier has fast, clean response (i.e., short rise times and no-after pulsing) and that noise pulses are not correlated with each other. If these requirements are not satisfied, information will be lost, and spurious pulses may be added by the discriminator. If the dark current does not follow Poisson statistics, false correlations will be introduced. Of the small number of photomultipliers that have been carefully evaluated and publicized, only a selected ITT FW 130 tube, labeled specially certified by the manufacturer, meets all the requirements reasonably well. Since this tube has an extended red response, it has a higher noise than many photocathode materials adequate for work with blue or green wavelengths. This, however, is compensated for by reducing the diameter of the photocathode. If some other tube is used, it is important to test the tube operation on well characterized scattering systems to establish that it has acceptable characteristics.

The scattering chamber consists of the sample holder and the associated optical components. For most work, temperature control of the sample must be included (at least to $\pm$ 0.1°). If only a small amount of the sample is available necessitating use of a small volume scattering cell, the cell should be surrounded with a fluid such as decahydronapthalene which matches the index of refraction to that of the scattering cell. In this way the scatter caused from the change in refractive index at the air/glass interface can be substantially removed from the vicinity of the scattering volume.

The collimated beam should be focussed at the center of the scattering cell. By reducing the diameter of the beam the coherence area of the scattered light at the photodetector is increased; that is the area of the photocathode which can contribute to the detected signal strength is enlarged. There are two common scattering cell geometries in use. The cylindrical cell is useful for measuring with scattering angles between 20° and 160°. Since the incident laser beam is so small, the radius of curvature of the scattering cell causes insignificant distortion of the scattering volume if the diameter of the sample container is greater than a centimeter. The rectangular cell is particularly useful for low angle scattering; however, it can be used for a range of larger angles. Corrections must be made for refraction for scattering at all angles except 90°. After the light leaves the cell, a lens and a baffle system are incorporated to define the length of the scattering volume and the uncertainty associated with the scattering angle and to prevent stray light from reaching the photomultiplier tube.

If the height of the incident beam has been altered with mirrors or prisms before it gets to the scattering cell, the state of polarization has probably been slightly changed. For simple QLLS analysis this is not significant, but if total intensity is simultaneously being monitored or if depolarization measurements are being taken, the beam must be repolarized.

Finally attention must be paid to the laser light source. Commercially available gas lasers are quite adequate. It is advantageous to have a laser tuned to the TEM_{00} mode which means that the light intensity possesses a Gaussian intensity profile, and the effective beam diameter has a minimum size. If the laser is not mode-stabilized, it will simultaneously produce several lines that are separated on the order of at least 10 MHz. Since this separation is much larger than the broadening of the Rayleigh line width, it is not necessary to use a mode-stabilized laser. Two lasers are in common use. The He–Ne laser produces a line at 633 nm and typically has a power of less than 10 mW. It's advantage is that it is relatively inexpensive. Alternately the argon ion laser, costing more, has 100 times more power in either of its two strongest lines at

488 and 514 nm where photomultipliers are substantially more efficient at detecting radiation. False noise spectra have been detected in Ar^+ lasers, however, when the plasma pressure is low (*13, 24*). Other problems have been reported concerning the mechanical stability of these laser tubes, but with improved cathode materials and provided the appropriate plasma tube pressure is maintained, these problems have vanished. In conclusion, many different types of spectrometer configurations can be constructed to gather QLLS data; for many investigations significant information can be obtained from an instrument that is relatively inexpensive.

Sample Preparation and Data Analysis

Preparation of optically clear samples is crucially important to obtain precise data, although in a few situations parasitic scattering from large dust particles has been used as a local oscillator source for heterodyne scattering. When dealing with laser light, the experimentalist has the advantage that since the light is more intense, a smaller volume of sample can be used. The degree of sample preparation depends substantially on the nature of the sample being studied. For most synthetic nonionic polymers, filtration of the sample through a membrane filter that is impervious to the solvent is sufficient. Pores should be large enough that the concentration of the sample is not altered substantially. Polyelectrolyte solutions, especially those with high charge density in solvents of low ionic strength, are more likely to tend to accumulate particulate material and in addition are often poorer light scatterers. They provide a greater challenge. In situ centrifugation of the filtered sample in the scattering cell immediately prior to insertion in the light scattering spectrometer is often the best way to purge these systems of large parasitic scatterers.

The data from a population of identical particles can be easily analyzed. If the homodyne autocorrelation function of monodisperse spheres (Equation 12) is measured, the data should conform to the expansion:

$$G^{(2)}(\tau) = A(1 + \beta|g^{(1)}(\tau)|^2)$$

where A is the background signal level and β is a value near 1 that accounts for nonideal coherence of the collected light. Generally β is considered an unknown parameter to be determined for each spectrum even though it should be a constant for the spectrometer. With photon counting equipment A can be determined with far less uncertainty than with analog equipment (Ref. *13*, Ch. 6). Since:

$$\ln(G^{(2)}(\tau) - A) = \ln A\beta + 2\ln|g^{(1)}(\tau)|$$
$$= \ln A\beta - 2D_tq^2\tau \quad (25)$$

a plot of the *LHS* vs. τ gives a straight line of slope $2D_tq^2$. If the power spectrum is recorded, Equation 21 appears on a baseline B' in the form:

$$P(\omega) = \frac{A'}{\omega^2 + (2D_tq^2)^2} + B' \quad (26)$$

where A' contains the normalization constants. This can be written as:

$$\frac{1}{P(\omega) - B'} = \frac{(2D_tq^2)^2}{A'} + \frac{\omega^2}{A'} \quad (27)$$

so a plot of $(P(\omega) - B')^{-1}$ as a function of ω^2 gives a straight line. The square root of the ratio of the y-intercept to the slope is equal to $2D_tq^2$. When least square procedures are used to determine D_t from either the correlation function or from the power spectrum, it is important to apply appropriate weighting of the data points to maintain the proper significance for each value after it is transformed (*25, 26*).

When there are more than one species present in a sample, each contributes to give a correlation function which is a sum of exponentials and a power spectrum which is a sum of Lorentzians. The intensity of each component in the composite function is proportional to the product of the molecular weight and the concentration in (w/v) units, assuming the refractive index increments of each component are identical. Two basic approaches are available to extract the particle distribution from the QLLS data.

The first approach assumes that the sample is composed of a discrete number or a specific distribution of components. Then the deviations from linearity for the proposed model can be included as higher order corrections in Equations 25 or 27. Current techniques are limited to the consideration of two discrete size populations of particles or to a size profile defined by two parameters such as the Schulz–Zimm distribution (*7, 26, 27, 28, 29, 30*). If there is evidence that such a model describes the sample, it is certainly the best way to proceed. This approach has been applied recently to polydispersity analysis based on the assumption of a particular distribution model by Chen et al. (*35, 36*) and McDonnell et al. (*26, 37*).

The alternative is the method of cumulants (*31*) which considers an unspecified size distribution of diffusing particles corresponding to a distribution $(G\Gamma)$ of relaxation time constants $\Gamma = D_tq^2$ which is normalized so that:

$$\int_0^\infty G(\Gamma)\,d\Gamma = 1 \tag{28}$$

Then the correlation function is:

$$g^1(\tau) = \int_0^\infty G(\Gamma)\,e^{-\Gamma\tau}d\Gamma \tag{29}$$

and the mean relaxation rate is:

$$\overline{\Gamma} = \int_0^\infty G(\Gamma)\,\Gamma d\Gamma \tag{30}$$

If the exponential term in the Equation 29 is rewritten as (*31*):

$$e^{-\overline{\Gamma}\tau}e^{-(\Gamma-\overline{\Gamma})\tau} \tag{31}$$

the correlation function can be expanded in a Taylor series as:

$$|g^1(\tau)| = e^{-\overline{\Gamma}\tau}(1 + (\Gamma - \overline{\Gamma})^2\tau^2/2! + (\Gamma - \overline{\Gamma})^3\tau^3/3! + \ldots \tag{32}$$

This procedure leads (*25, 31*) to a polynomial expansion of $\ln g^{(1)}(\tau)$:

$$\ln g^{(1)}(\tau) = -\overline{\Gamma}\tau + \frac{1}{2!}\frac{K_2}{\Gamma^2}(\overline{\Gamma}\tau)^2 - \frac{1}{3!}\left(\frac{K_3}{\Gamma^3}\right)(\overline{\Gamma}\tau)^3 + \ldots \tag{33}$$

where the zeroth moment $\overline{\Gamma}$ can be related to the z-average diffusion coefficient $<D_t>_z$:

$$\overline{\Gamma} = <D_t>_z q^2 \tag{34}$$

and

$$K_n = \int (\Gamma - \overline{\Gamma})^n G(\Gamma)\,d\Gamma$$

is the n'th moment about the mean $\overline{\Gamma}$. The second moment K_2 gives information on the width of the distribution $G(\Gamma)$; K_3 and K_4 depend on the asymmetry of the distribution. Unfortunately, these methods are of limited usefulness for analyzing molecular weight distribution. It is difficult to determine more than the second moment K_2 accurately, and this can be done only with difficulty for even moderately polydisperse polymers ($M_z|M_w \lesssim 2.0$ (*25*). The effects of intermacromolecular interactions and intramolecular relaxational components together with the associated changes in scattering amplitudes of larger molecular weight components as the scattering angle varies are additional problems.

Tanaka (32) and Brehm and Bloomfield (33) have discussed how to obtain additional information by studying the angle dependence of the mean Γ and the second moment K_2. Figures 2 and 3 illustrate data taken on our QLLS spectrometer for polydisperse samples of polystyrene (PS) in cyclohexane at a temperature slightly above the theta temperature and at a concentration of 10 mg/mL. Qualitatively, these results agree with the analysis of Tanaka (32). Each sample has approximately the same M_w, and the spectrum of the more polydisperse sample is narrower than that of the other. The extrapolated values for the average D_t are substantially different, as are the limiting slopes of the angle dependence. The magnitude of the differences observed in our data are somewhat larger than would be predicted by the theory of Tanaka (32) or by the earlier analysis of Frederick et al. (28) based on use of a Schulz–Zimm distribution. The significant angular dependence implies substantial scattering contributions from very high molecular weight species. We obtained limiting values of $<D_t>_z = 2.25 \times 10^{-7}$ cm^2/sec for sample 1 and $<D_t>_z = 1.95 \times 10^{-7}$ cm^2/sec for sample 2. These are comparable with values $D_t = 2.16 \times 10^{-7}$ cm^2/sec deduced by Frederick (28) for a monodisperse polystyrene sample of $M_w = 257{,}800$ at a concentration of

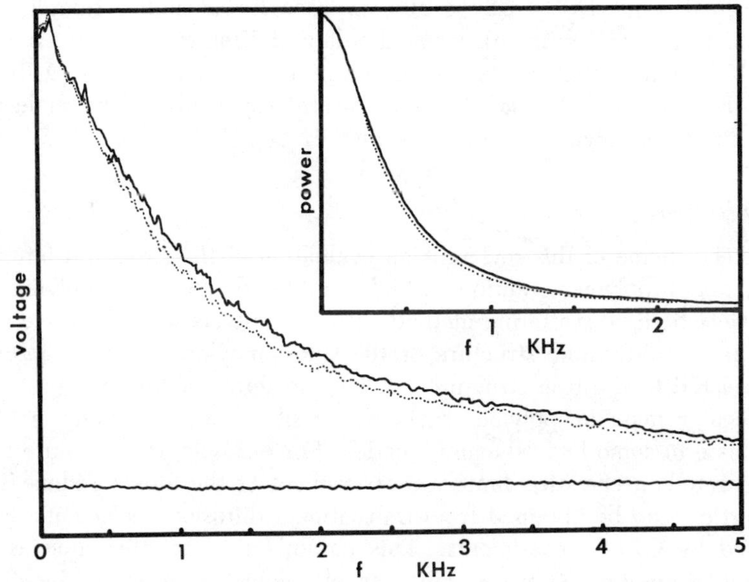

Figure 2. Voltage spectra of photocurrent produced by light scattered from solutions of polydisperse PS in cyclohexane at 37°C. Concentration = 10 mg/mL. Scattering angle θ = 40°. (——) ≡ sample with $M_n = 100{,}000$, $M_w = 250{,}000$. *(· · ·) ≡ sample with* $M_n = 59{,}000$, $M_w = 269{,}000$. *Inset are the corresponding power spectra (squared voltage spectra).*

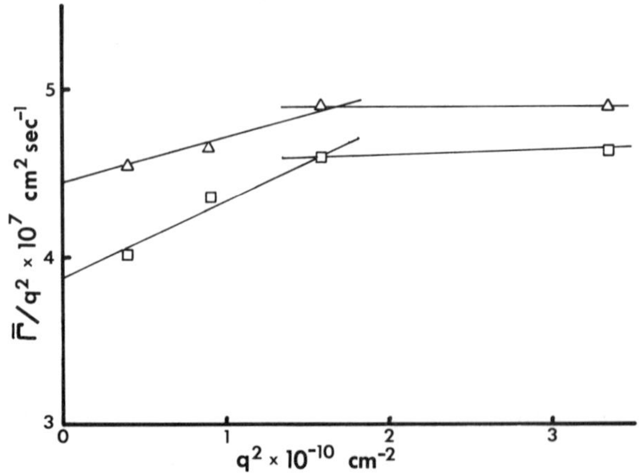

Figure 3. Effective translational diffusion coefficient $D_{eff} = \bar{\Gamma}/q^2$ *of polydisperse PS's in cyclohexane at 37°C.* (△) ≡ *sample with* $\bar{M}_n = 100{,}000$, $\bar{M}_w = 250{,}000$. (□) ≡ *sample with* $\bar{M}_n = 59{,}000$, $\bar{M}_w = 269{,}000$.

5 mg/mL, and $D_t = 2.2 \times 10^{-7}$ cm^2/sec for a similar sample at a concentration of 10 mg/mL using the data of King et al. (*34*). Analysis of QLLS data according to the prescriptions of Tanaka (*32*) or Brehm and Bloomfield (*33*) may be quite useful especially if the molecular weight of the polymer or colloid is high, $M > 10^6$.

Applications

The theme of this review is an evaluation of the structural information about polymer solutions which can be derived by applying the quasielastic light scattering method. The parameters we will derive refer to the hydrodynamic structure of the solution which cannot rigorously be related to the true structure (defined in terms of the distribution of molecular mass or polymer segment density in the solution) without recourse to some hydrodynamic model. For example, in the dilute solution, zero-concentration limit, a measurement of the Stokes radius R_s of a particle can be obtained from translational diffusion coefficients determined by QLLS experiments. This parameter is not the same as the radius of gyration R_g of the same particles, which is measured by studying the angular distribution of intensity of scattered electromagnetic radiation. In the case of rigid particles, R_s will include bound solvent species which move with the particle when it undergoes translational diffusion; in the case of flexible coil particles R_s is also affected by the relative degree of binding or draining of solvent through the chains and also by the hydrodynamic interaction of neighboring chain segments. If a mathematical connection between R_s and R_g is to be made, a suitable hydrodynamic model must be developed (*38*). Likewise, when considering more concentrated solutions to obtain information about intermolecular structure, e.g., correlation lengths, or the dynamics of entanglements, an analytical connection must be made between the dynamics of the system and the distribution of segment density or interparticle structure.

In dealing with the structural properties of polymer solutions, it is convenient to adopt the conventional notion of three regimes of polymer solution behavior (*39*):

(1) The dilute solution regime characterized by separated macromolecular domains in which the macromolecules are believed to behave essentially like a gas of hard spheres;

(2) The semi-dilute regime in which overlap of the excluded volume of individual particles occurs, perhaps resulting from overlap of the spheres of influence of macromolecules having long range (electrostatic) forces or from actual physical interpenetration of chain molecules;

(3) The concentrated solution regime characterized by values of the chain segment density or monomer concentration ρ such that $\rho \approx 1$, i.e., the solvent is present as the minor component in terms of relative volume fraction.

It will also be convenient in most of the discussion to treat phenomena associated with rigid macromolecules separately from the more complex events encountered in solutions of flexible polymer.

Dilute Solution Properties. INFINITE DILUTION LIMIT. In general terms, in dilute solution the concentration dependence of the translational diffusion coefficient of a polymer species has been written conventionally as:

$$D_t = D_t^{\circ}(1 + (k_D - \bar{v}_2)c_2 + \ldots) \tag{35}$$

where k_D is a diffusion virial term which in principle can be computed by a suitable hydrodynamic theory, c_2 is the polymer concentration in mg/cc, and $\bar{v}_2$ is the partial specific volume of the macromolecular solute. The term $\bar{v}_2 c_2$ is a correction which appears because experimentally D_t is measured in a constant volume frame of reference, i.e., when the solute has a net flux in one direction, the solvent must have a flux in the opposite direction to preserve volume (*8*). Hydrodynamic and thermodynamic contributions to k_D are usually theoretically evaluated in a reference frame in which the solvent is at rest. The classical avenue to k_D is to separately evaluate the contributions from thermodynamic and hydrodynamic interactions to D_t (*40*):

$$D_t = \frac{1}{N_A f}\left(\frac{\partial \pi}{\partial c_2}\right)_{\mu} \tag{36}$$

where f is the translational frictional coefficient and $(\partial\pi/\partial c_2)_\mu$ is the osmotic compressibility. Thus:

$$D_t = D_t^o[1 + (2A_2M - k_f - \bar{v}_2)c_2] \tag{37}$$

where A_2 is the second osmotic virial coefficient defined by Equation 3, k_f is a friction virial coefficient which characterizes the concentration dependence in dilute solution of the frictional coefficient f:

$$f = f_o(1 + k_fc_2 + \ldots) \tag{38}$$

Obviously Equation 37 has been terminated at linear terms in the dependence of polymer concentration c_2.

The diffusion coefficient extrapolated to $c_2 = 0$ can be related to a primary structural parameter, the Stokes radius R_s through the equation:

$$D_t^o = \frac{kT}{f_o} = \frac{kT}{6\pi\eta_oR_s} \tag{39}$$

where η_o is the solvent viscosity. Since R_s includes all solvent species bound to the particle, measurement of D_t^o by itself cannot directly permit either a measurement of M or of the structure (i.e., mass distribution) of a macromolecular particle without knowledge of the polymer–solvent interaction or other independent experimental data.

If particles are known to be spherical in shape and nondeformable in the relatively weak flow fields associated with Brownian motion (this may be expected in the case of synthetic latex particles, many proteins, and viruses and probably also holds for certain emulsion particles with rigid ordered interfaces, the Stokes radius will closely correspond to the hard sphere radius R, related to R_g through $R_g^2 = 3/5\ R^2$ and may also be similar to that observed in the electron microscope R_{EM}. The value of R_s should, however, on detailed inspection be greater than the radii measured by the latter methods because it includes bound solvent molecules. The discrepancy can be used to estimate the degree of solvation δ_1 grams solvent/gram of the particle through the relation:

$$\frac{4}{3}\pi R_s^2 = \frac{M}{N_A}(\bar{v}_2 + \delta_1v_1^o) \tag{40}$$

where v_1^o is the specific volume of solvent, provided M is known from independent analysis. Alternatively δ_1 can be determined simply by subtracting the volume corresponding to R (deduced from R_g or R_{EM}) from that corresponding to R_s.

If the sample is polydisperse, it is possible to extract four z-averages of the polymer radius distribution, $<R^{-1}>$, $<R^{-2}>$, $<R>$ and $<R^2>$ using the analysis of Brehm and Bloomfield (*33*). The four averages are found from the limiting slopes and intercepts of plots of $\overline{\Gamma}/q^2$ and K^2/q^4 vs. q^2 where $\overline{\Gamma}$ and K^2 are the first and second cumulants of the cumulant expansion (Equation 33), and q is the scattering vector given by Equation 5. We have recently estimated $<D>_z$ (and therefore the average $<R^{-1}>$ for a suspension of the fluorocarbon microemulsion Fluosol FC-43 (The Green Cross Corp., Osaka, Japan (*41*).) The value obtained (0.43 μ) compares very favorably with a value calculated from the published size distribution of the microemulsion particles, measured in the electron microscope (*42*) (0.46 μ). From Equation 39, since $R_s \propto M^\gamma$, corresponding averages of the molecular weight distribution could be derived from the hydrodynamic data providing the parameter γ is known. Application of these ideas, however, does depend on an ability to determine precise values of the second moment K_2. This may not be possible for highly polydisperse samples (*25*). If only the first moment $\overline{\Gamma} = <D_t>_z\, q^2$ can be measured with precision, it is still possible to deduce $<R^{-1}>$, $<R>$, and $<R^2>$ or higher moments (or the corresponding molecular weight averages if γ is also available) from the angular dependence of K_1 alone if there is reason to believe that the distribution of particle sizes is adequately characterized by a two-parameter equation, such as the Gaussian, log-normal, or Schulz–Zimm functions. These analysis are only practical if the particle sizes are very large.

The Stokes radius relates D_t^o to the unsolvated molecular weight of the macromolecular particle by a relation of the type:

$$D_t^o = KM^{-\gamma} \tag{41}$$

where K, γ are empirical constants which depend on the polymer chain configuration and on the polymer–solvent interaction.

The form of Equation 41 has been established by QLLS spectroscopy for several systems:

polystyrene/2-butanonone (*37*, *43*): (25°C)	$K = (3.1 \pm 0.2) \times 10^{-4}$ cm²/sec $\gamma = 0.53 \pm 0.02$
polystyrene/tetrahydrofuran (*37*, *44*): (30°C)	$K = (3.0 \pm 0.4) \times 10^{-4}$ cm²/sec $\gamma = 0.55 \pm 0.01$
protein/6MGdnHCl—0.1M (*2*, *6*): mercaptoethanol (MSH) (25°C)	$K = (9.68 \pm 0.7) \times 10^{-5}$ cm²/sec $\gamma = 0.56 \pm 0.02$

All of these results agree with published intrinsic viscosity data for the identical system through the expected relation:

$$1 + a = 3\gamma \tag{42}$$

where a is the exponent in the Mark–Houwink equation. The relationship for protein random coils is particularly useful for characterizing quaternary structure in biological proteins (*45*). Comparison of the molecular weights of the proteins in 6*M* GdnHCl/0.1 MSH with that of proteins in their native forms permits the number of polypeptide chains in oligomeric protein structures to be determined.

The availability of convenient means to determine accurate values of D_t^o adds a new dimension to some of the more classical routes to structural analysis in polymer solutions. Several studies have been reported (*46, 47, 48*) in which weight-average sedimentation coefficient measured in the analytical ultracentrifuge is combined with z-average D_t^o measured by QLLS to give weight-average values for molecular weight M_w using the Svedberg equation:

$$M_w = \frac{<S^o>_w}{<D_t^o>_z} \cdot \frac{RT}{(1 - \bar{v}_2\rho_o)} \tag{43}$$

where ρ_o is the solvent density. An interesting study has been described by Koppel (*49*) who separated ribosomes in the analytical ultracentrifuge in a sucrose density gradient and measured the values of D_t directly inside the gradient on the separated fractions. Since the concentration dependence of S and D_t was negligible, he was able to determine S, D_t, and M for several different macromolecular species in a single experiment using an independent estimate of $\bar{v}_2$. A similar experiment was reported by Lowenstein and Birnboim who have constructed an instrument based on a modified preparative centrifuge in which S and D_t of sedimenting polymers in a capillary tube can apparently be measured simultaneously by a phototube set to longitudinally scan the capillary tube without requiring use of a sucrose gradient (*50*). Another recent experiment (*51*) describes the separation of mixtures of charged polyelectrolytes in a polyacrylamide gel column by disc electrophoresis and subsequent characterization of D_t of the fractions in situ by QLLS techniques.

Correlation of D_t^o with intrinsic viscosity $[\eta]$ also permits a determination of M for spherically symmetric polymers through the Mandelkern–Flory equation:

$$M[\eta] = \beta(D_t^o\eta_o/T)^{-3} \tag{44}$$

where β is an empirical constant which should, according to theory, be very insensitive to polymer–solvent interactions. Experimental data for various polymer–solvent combinations are plotted in log form in Figure 4. The best linear fit represents a value $\beta = 3.0 \times 10^{-27}$ when $[\eta]$ is described in units of ccg^{-1}, D_t° in cm^2sec^{-1}, T in °K, and η_o in poise. The special utility of Equation 44 arises when one deals with copolymeric species which are difficult to characterize accurately by conventional techniques because of the complex nature of preferential interactions of each monomeric component with solvent species. Price et al. (*52*) have described the application of Equation 44 to determination of molecular weight of multimolecular micelles formed by a two-block copolymer of

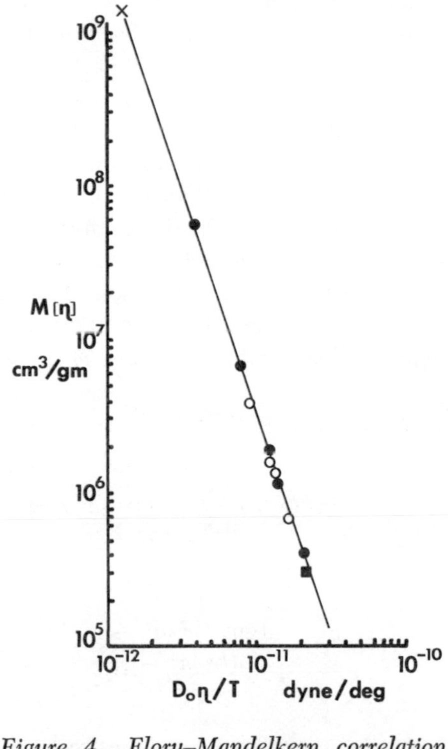

Figure 4. Flory–Mandelkern correlation of intrinsic viscosity, molecular weight, and translational diffusion coefficient for a variety of polymer solvent systems, demonstrating the insensitivity of these data to the structure of the macromolecule: (●), PS/tetrahydrofuran (37); (○), protein random coils in 6M guanidine hydrochloride–0.1M mercaptoethanol (26); (×), tobacco mosaic virus in aqueous solution (11); (■), bovine serum albumen in aqueous solution (37).

polystyrene–polyisoprene in n-decane. At 25°C the polymer exists predominantly in the form of micelles, with $M = 1.82 \times 10^6$ and Stokes radius $R_s = 196$Å; as the temperature is raised, the equilibrium shifts towards the monomeric species until at 65°C, the polymer is entirely monomeric, $M = 65{,}000$.

Equation 44 can be used to estimate the molecular geometry of rigid anisometric particles in terms of an equivalent ellipsoid of revolution since under these circumstances β is no longer constant but varies with axial ratio. Analysis of this type has been applied to synthetic polymers which exist in conformationally rigid forms (*53*) such as helices. Comparison of D_t° with $[\eta]$ for narrow fractions of the polymer of known molecular weight can furnish an estimate of the helical pitch (*53*), provided data for the axial ratio of each fraction are extrapolated to small values of M, to avoid the problems of partial flexibility at higher molecular weights. However, if particles are only moderately anisometric, Equation 44 is not very useful, and, in fact, varies by less than 5% for axial ratios less than 10. The rotational diffusion coefficient D_r° is much more sensitive to particle shape than D_t° or $[\eta]$, and, since accurate methods for measuring D_r are also available including transient birefringence, flow dichroism, dielectric relaxation, and fluorescence anisotropy as well as QLLS spectroscopy, comparison of D_t° and D_r° can be a very useful way to characterize particle shape. Perrins equations for the translational frictional coefficient take the form (*54*):

$$f = 6\pi\eta a/G(b/a) \tag{45}$$

where

prolate ellipsoid $a > b$:

$$G(b/a) = \frac{\ln\{(b/a)^{-1}[1 + (1 - b^2/a^2)^{\frac{1}{2}}]\}}{(1 - b^2/a^2)^{\frac{1}{2}}} \tag{46}$$

oblate ellipsoid $a < b$:

$$G(b/a) = \frac{\tan^{-1}(b^2/a^2 - 1)^{\frac{1}{2}}}{(b^2/a^2 - 1)^{\frac{1}{2}}} \tag{47}$$

Correspondingly the rotational coefficient is defined as:

$$f^R = \frac{16\pi\eta a^3}{3}\left\{\frac{1 - (b^2/a^2)^2}{[(2 - b^2/a^2)G(b/a) - 1]}\right\} \tag{48}$$

where $G(b/a)$ has the same meaning as in Equation 46 and 47. Methods are available for analytical (*55*) or numerical inversion (*56*) of Equations 45–48 which prove surprisingly useful in characterizing the dimensions of even small protein molecules. From the values $D_t^\circ = 10.6 \times 10^{-7}$

cm^2/sec and $D_r^o = 16.7 \times 10^6$ sec^{-1} for aqueous solutions of lysozyme determined by Dubin (*57*) using depolarized quasielastic light scattering analysis, one deduces an equivalent prolate ellipsoid with dimensions (*56*) $2a = 55$Å; $2b = 34$Å. Likewise, for the protein bovine serum albumin, $D_t^o = 6 \times 10^{-7}$ cm^2/sec measured by light scattering and $D_r^o = 1.93 \times 10^6$ sec^{-1} measured by transient birefringence or dielectric relaxation lead to an equivalent prolate ellipsoid (*56*) having $2a =$ 144Å and $2b = 38$Å.

The translational diffusion coefficient can be used to follow conformational changes in macromolecules which result in changes in their hydrodynamic behavior. Several studies of the helix-to-coil transitions of synthetic polypeptides (*58, 59, 60*) in which the R_s of the helical form is much larger than that of the coil. Studies of the helix-to-coil transition of poly-L-lysine HBr by QLLS (*59, 60*) have confirmed earlier viscometric studies that the partially helical state of the polymer intermediate between the fully helical and random forms has a smaller R_s than either of the extreme forms, indicating the partially α-helical form has a more compact structure. The random chain is contracted by the formation of short helical segments prior to long range extension of the helical conformation. The real advantage of the QLLS method, however, lies in the study of more subtle changes in conformational organization. With current photon-counting systems, it is possible to measure D_t and therefore R_s^{app} to within 1% accuracy. Recently, Chu (*61*) has detected various discrete conformational states in the thermal denaturation of t-RNA corresponding to changes of only a few percent in D_t.

At another level, the convenience and accuracy of D_t^o measurements by QLLS provides a useful addition to the experimental arsenal for elucidating details of the unperturbed dimensions of flexible coil polymers in solution. These can be measured directly by studying the translational diffusion of polymer in a theta solvent, using the theoretical expression:

$$D_t^o = kT/P_o\eta_o\langle R_g^2\rangle^{\frac{1}{2}} \tag{49}$$

where $P_o = 5.1$ according to the Kirkwood–Riseman theory (*62*). For a nondraining, unperturbed coil:

$$\langle R_g^2\rangle^o = (1/6)na^2 \tag{50}$$

and n is the number of chain segments, and a is the unperturbed effective bond length which is determined by the short range interactions of neighboring monomers and can be theoretically evaluated using conformational energy calculations. In good solvents, the chain dimensions expand because of the polymer–solvent interaction:

$$<R_g^2> = \alpha_r^2 <R_g^2>_o \quad (51)$$

where α_r is a chain expansion parameter. The polymer–solvent interaction also effects the hydrodynamic interaction, and therefore the parameter P_o in Equation 49 is also modified. In such a case, the unperturbed dimensions must be obtained by an extrapolation procedure (*63*).

For example, one can define a chain expansion parameter α_f which characterizes the effect of excluded volume on the Stokes radius R_s by:

$$\alpha_f = R_s/R_s^o = kT/[(6n)^{\frac{1}{2}}\pi\eta\xi a D_t^o] \quad (52)$$

where

$$\xi = R_s/R_g$$

and

where R_s^o is the Stokes radius in an ideal solvent. Stockmayer and Albrecht (*63*) calculate that α_f can be characterized by a series perturbation expansion in the excluded volume parameter z:

$$\alpha_f = 1 + 0.609z - \ldots \quad (53)$$

where

$$z = (3/2\pi)^{3/2} v n^2 / <L^2>_\theta^{3/2} = (3/2\pi)^{3/2} v n^{\frac{1}{2}} / a^3 \quad (54)$$

v is the effective volume of a hydrodynamic subunit, and $<L^2>_\theta = na^2$ is the unperturbed root-mean square end-to-end distance. Equations 50–54 lead to the relation:

$$(D_t^o n^{\frac{1}{2}})^{-1} = (6^{\frac{1}{2}}\pi\eta\xi a/kT)[1 + 0.201(v/a^3)n^{\frac{1}{2}} - \ldots] \quad (55)$$

which predicts a linear relation between $(D_t^o n^{\frac{1}{2}})^{-1}$ and $n^{\frac{1}{2}}$ at low z or $n^{\frac{1}{2}}$. We have applied this analysis to diffusion data for denatured proteins in 6*M* GdnHCl − 0.1 *M* MSH (*26*) and determined a = 9.7Å, and to polystyrene fractions in tetrahydrofuran with the result a = 5.6Å. The latter value was determined based on an analysis which includes higher terms in Equation 53 (*37*). Each of these values for a is in close accord with earlier experimental and theoretical studies. The advantages of QLLS for studies of unperturbed dimensions and the hydrodynamic theories embodied in Equations 50 and 53, compared with more conventional methods, are convenience and accuracy. None of the experimental investigations of this type reported to date using the QLLS method approaches the limits of precision of modern photon-counting techniques (*13*). A further special utility of diffusion measurements for studies of unperturbed dimensions in comparison with intrinsic viscosity analyses, is that α_f in Equation 53 is negligibly different from α_r in Equation 51 which is accessible by conformational energy calculations (*43, 63*). This is not true for the corresponding quantity α_η defined by:

$$\alpha_\eta{}^3 = [\eta]/[\eta]_\theta \tag{56}$$

In summary, the previous section indicates that in the condition of infinite dilution, the hydrodynamic theory of translational diffusion is quite well understood and permits one to develop useful structural information about the dissolved macromolecules or suspended colloid from $D_2{}^o$ either by itself or combined with other hydrodynamic parameters. The limiting diffusion coefficient of rigid particles approximates the Stokes–Einstein–Perrin model of impermeable ellipsoids, and the diffusion of flexible coils is apparently described quite accurately by the Kirkwood–Riseman theory (*62*) in theta solvents where the intramolecular excluded volume contribution is zero. The effects of finite excluded volume are quite well explained by the two-parameter theory (*38*) at least in cases where the excluded volume effect is small. We can, however, point to some apparent experimental discrepancies which require incorporation of more subtle features of the hydrodynamic structure. Thus, experimental values of the β parameter in the Mandelkern–Scheraga Equation 44 measured for certain globular proteins and micelles are smaller than the limiting value for impermeable rigid spheres. Recently McCammon et al. (*65*) have suggested that the explanation for this requires incorporating finite porosity of the particles into the theoretical formalism. Likewise, the β parameter measured for flexible coils in some experimental reports (*26, 37, 44*) seems to be significantly smaller than should be expected on the basis of the Kirkwood–Riseman theory. Also, the theoretical analysis of the limiting hydrodynamic parameters $D_t{}^o$ and $[\eta]$ under conditions of large excluded volume is not well understood. Diffusion studies using QLLS spectroscopy can provide a fruitful approach to establishing an accurate experimental data set for exploring these theoretical problems.

When dealing with large rigid anisometric macromolecules, rotational diffusion introduces additional Lorentzian spectral components (or exponential terms in the correlation function) with halfwidth (time constant)

$$\Gamma_{rot} = D_t q^2 + (6D_r)^{-1} \tag{57}$$

Thus by studying the angular dependence of the linewidth of the isotropic component at large angles, it is possible to deduce D_r. This parameter can also be obtained if the particles are optically anisotropic by studying the depolarized light scattering at forward scattering angles where $D\kappa^2$ is negligible compared with $(6D_r)^{-1}$. Rotational diffusion coefficients of tobacco mosaic virus have been determined by both of these techniques (*66, 67, 68, 69*), and for lyzozyme (*57*) by depolarized scattering which agrees well with measurements by other relaxation techniques.

Finally, when studying dilute solutions of high-molecular-weight flexible coil macromolecules, an additional contribution to the spectral dispersion of the scattered light can arise from the dynamic behavior of the low frequency, long wavelength internal vibrational motions of the chains (*70*). Additional Lorentzian spectral components (or exponential components of the correlation function) arise through this mechanism characterized by halfwidth (time constant):

$$\Gamma_{\text{vib}} = D\kappa^2 + \tau_i^{-1} \tag{58}$$

where τ_i^{-1} are the relaxation times of the internal modes. These are usually described by the Rouse–Bueche–Zimm bead-spring model (*71*). Like the rotational diffusion, the internal modes (primarily the longest wavelength, slowest component) can be most conveniently detected either from an additional linewidth increment in the polarized isotropic scattering which becomes significant at wide angles (*72*) or, if the molecules are optically anisotropic, by measuring the linewidth of the depolarized scattering at small angles where $D\kappa^2$ is usually negligible compared with τ_1^{-1} (*73*). Schurr (*74*) has pointed out that the latter procedure yields τ_1^{-1} directly, the former yields $1/2\tau_1$. This is equivalent to stating that the polarized wide-angle experiment measures the relaxation of the end-to-end distance while the depolarized small angle experiment measures the relaxation of the square of the end-to-end distances, the latter being the phenomenon also detected in dynamic viscosity and stopped-flow dichroism. The results, $\tau_1 = 1.8$ msec for PS (*72*) with $M = 27.3 \times 10^6$ in 2-butanone at 25°C and $\tau_1 = 18$ msec for calf-thymus DNA (*73*) with $M = 15 \times 10^6$, obtained by QLLS are in excellent agrement with values obtained by other relaxation methods such as dynamic viscosity and stopped flow dichroism.

Concentration Dependence of D_t. The concentration dependence of the frictional coefficient of macromolecules in dilute solution is another aspect which has recently been the subject of theoretical and experimental activity and which is quite sensitive to the details of hydrodynamic structure of individual particles.

In classical analysis, the concentration dependence of D_t has been described by Equations 35–37. For the simplest hydrodynamic model, that of impermeable neutral hard spheres, Pyun and Fixman (*75*) derived the result:

$$k_f = k_f^{\Phi} \frac{N_A V_h}{M} \tag{59}$$

where V_h is the hydrodynamic volume of the polymer which may be estimated from the Stokes radius (frictional coefficient) or intrinsic viscosity. The constant k_f^{Φ} takes the value 6.16 in the hard sphere calculation (*75*). Since the hydrodynamic radii measured by $[\eta]$ or f_o are not the same for flexible macromolecules or for molecules with a degree of porosity, k_f is not well defined for such species within the Pyun–Fixman theory. Usinc $\bar{v} \sim N_A \overline{V}_h/M$, we obtain from Equations 37 and 59:

$$k_D - \bar{v} = 2A_2M - 7.16N_AV_h/M \qquad (60)$$

When the neutral hard sphere result for $2A_2M$ is incorporated, the conclusion is that $k_D - \bar{v}$ is very small:

$$(k_D - \bar{v}) = -0.84N_AV_h/M \qquad (61)$$

Phillies (*76, 77*) has subsequently discussed the empirical extension of this classical diffusion theory to higher terms in the virial expansions. Recently, Batchelor has carried out a more detailed analysis of the intermolecular hydrodynamic interactions contributing to k_f and arrived at the conclusion (*78*):

$$(D_t/D_t^o) = (1 + 1.34\phi) \qquad (62)$$

where ϕ is the volume fraction of macromolecular particles. Adopting the suggestion of Alpert (*80*) that ϕ should be based on the solute hydrodynamic volume, Equation 62 corresponds to:

$$(k_D - \bar{v}) = 1.34N_AV_h/M \qquad (63)$$

On the other hand, a theoretical analysis by Anderson and Reed (*79*) concludes that for neutral hard-sphere molecules:

$$(D_t/D_t^o) = 1 - 1.83\phi \qquad (64)$$

or

$$(k_D - \bar{v}) = -1.83N_AV_h|M \qquad (65)$$

This derivation differs significantly from the classical approach embodied in the work of Phillies (*76, 77*) and Batchelor (*78*) in that the thermodynamic and frictional forces between interacting particles are not separately averaged over particle configurations (*79*).

There is a distinct lack of experimental evidence with which to compare these equations. Equation 65 closely predicts the concentration dependence of D_t of the proteins bovine serum albumin (BSA) and methemoglobin measured by Keller et al. (*81*) using classical diaphragm diffusion cell techniques (*79, 80*). Another prediction of the Anderson–Reed theory—that the mutual diffusion coefficient D_t measured by concentration gradient experiments is identical to the tracer diffusion coefficient $D_{t,T}$ determined by tracer diffusion techniques—also agrees with the experiments of Keller et al. (*81*). This evidence of course does not agree with the classical diffusion theory, and Phillies and co-workers (*22*) have reported BSA diffusion coefficients obtained using QLLS techniques which are substantially different from those of Keller et al. and interpreted as supporting the classical theory embodied in Equation 37. Obviously further accurate diffusion studies on model systems are needed to resolve these questions.

In comparing diffusion coefficients of charged colloidal particles in fluids at finite concentration with hard-sphere diffusion theories, it is important to take proper account of long range Coulombic repulsions between particles which may be significant even at relatively high ionic strengths. These forces will be absent only at the isoelectric point where the net charge is zero. For example, the comparison made by Phillies (*22*) of BSA diffusion coefficients in 0.2*M* NaCl at the isoionic pH 5.0 with Keller's data at the isoelectric pH 4.7 in 0.1*M* acetate buffer is invalid. Because of specific adsorption of chloride ion, BSA may have a net charge of $\bar{z} \approx -10$ e.u. in the former system (*82*). Phillies data at pH 4.7–4.9 and 0.2*M* NaCl exhibit a significant negative concentration dependence more in accord with Kellers data (*81*) and the Anderson–Reed theory (*79*).

We have recently carried out a study of BSA diffusion in two solvent systems which underline the significant effect that protein charge can have despite the presence of high salt concentrations. We obtained the experimental relations:

(i) pH 4.7, 0.1*M* acetate, $T = 22°C$:

$$(D_t/D_t^{\circ}) = 1 - 1.45(N_A V_h/M)_c, \text{ and} \tag{66}$$

(ii) pH 7.4, 0.15*M* NaCl, $T = 22°C$:

$$(D_t/D_t^{\circ}) = 1 + 1.34(N_A V_h/M)_c \tag{67}$$

These results are qualitatively in accord with the analysis of Anderson and Reed (*79*). System (i) is at the isoelectric point, thus $\bar{z} = 0$, and a neutral hard-sphere model should apply. Obviously the experimental data in Equation 66 are in reasonable agreement with Equation 65 but not with Equation 61 or 63. For system (ii), we calculate a net charge on the protein $\bar{z} = -20$ e.u. Using the Anderson–Reed theory with a screened Coulombic repulsion interaction force corresponding to this charge, we estimate:

$$(D_t/D_t^{o}) = 1 + 0.8(N_A V_h/M)_c \tag{68}$$

The two theoretical viewpoints discussed above are qualitatively in agreement in predicting the effect of long range forces on the diffusion of hard spheres; repulsive forces enhance positively the concentration dependence of D_t and $D_{t,T}$ over the neutral hard sphere result; attractive potentials result in an enhanced negative contribution. Important points which remain to be understood in interpreting studies of diffusion of proteins at finite concentrations when $\bar{z} = 0$ are

(1) The role of contact interactions

(2) The effect of nonspherical geometry

(3) The observations noted by several authors (*22, 64*) that diffusion coefficients D_t measured by QLLS become systematically larger than those measured by diaphragm cell techniques (*81*) at high concentrations.

Corresponding theoretical analysis of concentration dependence of D_t in solutions of flexible coil macromolecules have been carried out exclusively using the classical theory embodied in Equation 37 and have also turned out to be very sensitive to details of the particular model. The original treatment of Pyun and Fixman (*75*) based on a model of soft interpenetrable spheres of uniform segment density has been most commonly applied to these data. Their result is:

$$k_f = [7.16 - K(A)] \frac{N_A V_h}{M} \tag{69}$$

where $K(A)$ is a monotonically decreasing function of A which is related to A_2, the second osmotic virial coefficient. At the theta temperature $A = 0$, and $K(O) = 4.20$, 4.56, or 4.93, depending on three approximations for the binary clusters of overlapping spheres. The central value leads to:

$$k_f^{\theta} + \bar{v} = 2.6 \frac{N_A V_h}{M} \tag{70}$$

Yamakawa (*38*) and Imai (*83*) have published an alternative description based on a random coil model and the Kirkwood–Riseman theory (*62*) and obtained for theta-solvent conditions an equation equivalent to:

$$k_f^{\theta} + \bar{v} = \frac{N_A V_h}{M} \tag{71}$$

Most recently, Freed (*84*) has developed a generalized theory in terms of a multiple scattering representation of scattering of fluid waves by the polymer molecules and derived the precise result:

$$k_f^{\theta} + \bar{v} = [\eta] \equiv \nu N_A V_h / M \tag{72}$$

Equations 70 and 72 are numerically very similar using the value ν = 2.5 for any particle with spherical symmetry and have been reported to accurately describe k_f of PS and poly-α-methylstyrene in cyclohexane at the theta temperature (*84*) based on sedimentation data. However, recent translational diffusion measurements of polystyrene/cyclohexane solutions under theta conditions using QLLS indicate experimental k_f values which lie between the extremes represented by Equation 71 on one hand and Equations 70 and 72 on the other (*34*). For smaller molecular weights, the values are closer to the Pyun–Fixman or Freed theory; for high molecular weights, they are closer to the Yamakawa–Imai result.

Caroline and co-workers have recently reported measurements of translational diffusion coefficients in solutions of PS in two mixed-solvent systems at or near theta conditions. In the solvent CCl_4–methanol (*85*), they observed the "diffusion theta" state, defined when the coefficient γ of Equation 41 equals 0.5, to occur at 25°C and a volume fraction of CCl_4, $\phi CCl_4 = 0.8025$. In this system there is strong preferential adsorption of the polymer for CCl_4, and it is not possible to define a true theta state such that $\gamma = a = 1/2$ and $A_2 = 0$ simultaneously. Under diffusion theta conditions, the concentration dependence of D_t apparently is closely described by the Pyun–Fixman hard-sphere model. In the mixed solvent benzene–2 propanol, polystyrene exhibits a true theta condition at T = 25.5°C and ϕ (benzene) = 0.04. Frost and Caroline confirmed that γ = 0.5 within experimental error in this system (*86*) and report that values of the parameter k_f^{θ} are scattered between the extreme values corresponding to the predictions of Yamakawa (and Imai) and the soft-sphere model of Pyun–Fixman (or the Freed theory).

When flexible coils are dissolved in good solvents, the theoretical formulation of k_D is more difficult. The Pyun–Fixman theory represented by Equation 69 indicates k_f^{Φ} should increase monotonically from a value around 2.23, corresponding to theta solvents, to the hard sphere value 7.16, as the excluded volume effect increases. Yamakawa notes (*38*) that the Pyun–Fixman equation can be put in the approximate form:

$$k_f + \bar{v} = 0.2 k_f^{\Phi} [\eta] \tag{73}$$

while his own analysis based on the flexible coil model can be described by:

$$k_f + \overline{v} = 1.2A_2M - 0.2[\eta] \tag{74}$$

Equation 73 satisfactorily explains frictional coefficient data for PS in a variety of solvents (*87*). The thermodynamic contributions to k_D are modelled quite well using the two-parameter theory of polymer solutions (*87*). Recent QLLS experiments on PS/tetrahydrofuran solutions are also in accord with the Pyun–Fixman theory (*37, 44*). However, other studies of the frictional coefficient of poly-α-methylstyrene in *trans*-decalin and toluene indicate that the Pyun–Fixman theory but not the Yamakawa theory fits the data at small excluded volumes (*88*). Neither theory works well at large excluded volumes (*88*).

An important potential application of the QLLS method to characterization of protein solutions stems from the fact, discussed above, that k_D is quite small for solutions of rigid spheres. In systems of this type which exhibit substantial self-association, the major contribution to k_D^{app}, the apparent concentration dependence of D_t, derives from the association process which contributes a large negative term to the second virial coefficient. Herbert and Carlson (*89*) have reported a study of this type on the dimerization of the muscle protein myosin which indicates a value $K = 1.30$ dL/g for the association constant at lower concentrations of phosphate or sulfate (0.2*M*) and K = 10.6 dL/g at high concentrations (0.5*M*) (*89*). Chu et al. (*90*) have discussed some of the instrumental limitations of such analysis.

Dilute Solutions of Strongly Interacting Macromolecules. We intend to include by this definition those polymer solutions in which the range of interparticle interactions is comparable with the mean interparticle distance. The polymer particles are still believed to interact essentially by means of hard sphere type forces, but they can execute independent Brownian movements only to a limited degree since the solutions are crowded in the sense that the intermolecular interactions inevitably perturb the motion of the particle soon after it initiates a translational displacement. The situation can be brought about either by an increase in particle size and/or number density, which leads to an increase in the repulsive excluded volume force, or by the presence of a long range potential, which can be either repulsive (electrostatic) or attractive (electrostatic or electrodynamic, dipolar, van der Waals, hydrogen bonding, etc.). When the macromolecular dimensions and range of interactions are comparable with the wavelength of scattered light, unusual nonlinear behavior of the angle dependence of scattered intensity is observed which arises essentially because of interparticle as well as intraparticle interference effects. At the same time, nonlinear angular dependence of the linewidth of the scattered light is apparent (*91, 92, 93, 94, 95*). These authors have also noted the nonexponential or non-Lorentzian nature of the dynamic light scattering data. In some cases it is apparent that two distinct dynamic modes of fluctuation can be discerned, a fast and a slow mode, each of which has the characteristics of a translational diffusion process (*91, 94, 95*). Recent progress has been made in explaining these features of the light scattering properties.

The simplest system in terms of mathematical analysis is that of a suspension of hard spheres interacting by means of a long range Coulombic repulsion force. Data are extant for a suspension of the highly charged virus R17 in water at low ionic strength (*92, 93*), and for polystyrene latex spheres in water at low ionic strength (*94*). The diffusion coefficient D_t determined by a single Lorentzian fit to the spectral distribution of scattered light through Equation 18 or single exponential fit to the correlation function using Equation 20 should be a unique quantity, independent of scattering vector q, if the particles are monodisperse and noninteracting. Berne and Schaefer (*93*) found, however, that D_{eff}, the effective diffusion coefficient defined by $D_{eff} = \Gamma/q^2$, for a suspension of highly charged R17 virus particles at low ionic strength was not independent of scattering angle but was larger at small q-values and decreased in magnitude as q increased. These authors also noted the nonexponential nature of the experimental correlation functions and that D_{eff} was determined from the initial slope of the decay function. This behavior of D_{eff} was paralleled by an increase in the integrated scattered intensity from low values at small q to higher values at large q. In the presence of 1*M* NaCl, which screens the long range electrostatic forces, the classical diffusion behavior was observed, and an angle-independent decay constant was observed corresponding to the conventional diffusion constant D_t.

These phenomena are in accord with a theoretical analysis of the decay properties of the concentration fluctuations $\delta c(q,t)$ which formally includes the influence of long range intermacromolecular ordering forces on the diffusion of the particle. The result obtained by Berne and Schaefer (*93*) is:

$$\delta c(q,t) = \exp(-\Gamma(q)\tau) \tag{75}$$

where

$$\Gamma(q) = D(q)[S(q)]^{-1}q^2 \tag{76}$$

The kinetic factor $D(q)$ characterizes the average particle flux, and $S(q)$, the interparticle structure factor, is equivalent to the integrated scattered light intensity. When the range of interparticle interactions σ is comparable with d, the mean interparticle distance, then $D(q)$ may exhibit angle-dependent behavior. In the limits $q \to 0$, $d \to \infty$, $D(q)$ reduces to the translational diffusion constant D_t (*93*). The data of Berne and

Schaefer indicate that $S(q)$ is a decreasing function of q^2 as $q \rightarrow 0$ as expected for a weakly correlated repulsive interaction ($\sigma < d$) and that $D(q)$ depends weakly on q so that the effective diffusion coefficient, D_{eff}, ($D_{eff} = \Gamma(q)/q^2$) varies with q as $[S(q)]^{-1}$. They also note that, for more strongly correlated systems, $\sigma \sim d$, one should expect a broad maximum in $S(q)$ at $q > 2\pi$, analogous to the lowest order peak in the x-ray or neutron scattering structure factor of dense liquids. Also, for $\sigma >> d$, corresponding to long range order, $S(q)$ should exhibit sharp Bragg peaks at $q = 2n\pi$, where n is an integer. The latter situation appears to have been encountered by Brown et al. (*94*) in studies by photon correlation spectroscopy of the light scattering from dispersions of PS latex spheres 250Å in radius at low ionic strength. The angular dependence of the reciprocal of the effective diffusion coefficient determined from the initial slope of the autocorrelation decay function of the scattered light shows angular dependence similar to that of the mean scattered intensity corrected for $P(\theta)$ as predicted by Equation 76, this time showing a well defined maximum at intermediate q-values. At small q, $S(q)$ decreases below the noninteracting particle value of unity; in the large q limit, $S(q)$ tends to the limiting value unity, although, at higher concentrations, a second broad peak in $S(q)$ at larger q is evident, indicating the presence of a diffuse next nearest neighbor shell in addition to the nearest-neighbor shell responsible for the sharp primary peak. On dilution, these peaks are shifted to smaller q, taken to imply the interparticle structure expands as the number of particles is reduced.

An interesting feature of the correlation functions of the scattered light is that they are nonexponential to a degree which seems incompatible with a simple explanation in terms of sample polydispersity (*94*, *96*). The experimental data are in accord with a theoretical model (*94*, *96*) which decomposes the fluctuating velocity of a particle in an interacting system into a large-amplitude Brownian component which fluctuates rapidly and exhibits behavior identical to the short-time properties of the light scattering correlation function, Equation 73; and a small-amplitude slow component whose time constant is comparable with that necessary for the particle to move a significant fraction of the interparticle distance $d \sim q^{-1}$ and which apparently gives rise to a slow (low frequency) light scattering component with characteristic frequency (*96*):

$$\Gamma_{int} = D_{int} q^2 \qquad (77)$$

where D_{int} is a cooperative diffusion coefficient dependent on the interparticle interaction. A recent theoretical paper by Ackerson (*97*) shows how a cumulant analysis based on the generalized Smoluchowski equation (*98*) can be used, in principle, to model not only the way in which hydrodynamic interactions of the particles with solvent modify the short-time decay behavior of the concentration fluctuations, but also the longer-time decay characteristics corresponding to Γ_{int} where the interparticle potential is important. These light scattering phenomena are the analog of the narrowing in frequency of the quasielastic coherent neutron scattering spectrum $S(q,\omega)$ of neutrons scattered by simple liquids, when q takes a value near a maximum in the corresponding structure factor $S(q)$ (*99*). This phenomenon was theoretically predicted by de Gennes (*100*).

Unusual dynamic light scattering effects have also been observed in congested solutions of flexible coil macromolecules where intermolecular repulsive interactions should again manifest themselves. Jamieson and Presley (*91*) reported observation of fast and slow concentration fluctuation components in dilute solutions of a partially hydrolyzed polyacrylamide solution in which high molecular weight coupled to the high charge density on the polyion should lead to highly perturbed translated diffusive motions. At the time, it was suggested (*91*) that these phenomena corresponded to anisotropic diffusion behavior of nonspherical macromolecular coils which results from the intermacromolecular interactions. At wide angles, a slow spectral component was observed whose linewidth exhibited q^2-dependence and was ascribed to partially rotation-average diffusion perpendicular to the chain axis; at smaller angles a faster component was seen which also appeared to vary as q^2 and was proposed to result from diffusion parallel to the chain axis. Subsequently, the autocorrelation function of light scattered by certain DNA solutions was also found to exhibit fast and slow components (*73, 95, 101, 102*), each linearly dependent on q^2. The time constant of the fast component in DNA solutions is similar to that expected for translational diffusion with the diffusion coefficient D_t for unhindered motion; the relaxation time for the slow component is an order of magnitude slower than the normal diffusion process and is similar to that of the internal modes. The relaxation time for the slow mode increases rapidly with increase in molecular weight or in the presence of acridine orange which extends the persistence length of DNA (*103*). The amplitude of the slow mode, on the other hand, decreases substantially as DNA concentration is reduced or when salt is added (1*M* NaCl) (*104*).

The theoretical analysis of macromolecular diffusion in interacting solutions of flexible macromolecules is more difficult than that for hard spheres. The presence of internal fluctuations in chain configuration must lead to a softening of the interparticle force field and also means that the particle geometry may not be constant. Lee et al. (*104*) have recently published a theoretical analysis which describes the qualitative features of the dynamic light scattering data in DNA solutions. The hydrodynamic model used is based on the assumption that the instantaneous shapes of the DNA molecules are nonspherical (*105*) and that consequently, in congested solutions, there is a severe anisotropy of the translational diffusion. Diffusion is imagined to occur nearly normally along the long axis of the ellipsoidal DNA coil but is essentially zero in the perpendicular direction. Under these crowded conditions, it is plausible that the internal motions of the chains, acting to change the chain configuration and to reorient the long axes will be coupled strongly to the external translational diffusion (*104*). Solution of the equations of motion for this hydrodynamic model leads to the prediction of fast and slow modes for relaxation of the concentration fluctuations, the amplitude of the slow mode being relatively larger in agreement with the experimental observation (*104*). As $q \rightarrow 0$, all of the scattering amplitude appears in the slow mode whose time constant is characterized by an averaged diffusion coefficient. The analysis assumes that the internal mode relaxation behavior is unaffected by the intermolecular interaction and, further, neglects intraparticle interference contributions to the scattering intensity. The authors also show that there is no hydrodynamic mechanism by which independently orienting macromolecules could produce a relaxation component slower than that of translational diffusion. They further discount the possibility that hard-sphere interactions corresponding to long range order are the source of the anomalous scattering component.

Lee et al. (*104*) note that their theoretical analysis disagrees with the data of Jamieson et al. (*91*) in which the bulk of the relaxing amplitude apparently shifts into the fast mode at small angles and suggest that a "cage" phenomenon may be present in the latter solutions. We note there are qualitative similarities between the linewidth data for the ionized polyacrylamide system and the data for Berne and Schaefer for the charged R17 virus indicating that 'hard-sphere' repulsive interactions may dominate for this highly charged polyion. However, since these experiments (*91*) were performed in a heterodyne configuration, the question of relative amplitudes of the relaxation modes at small q-values in this system is open and is being investigated further.

Experimental and theoretical studies of dynamic light scattering from fluid mixtures in which there are strongly attractive long range intermolecular forces between the components have been reported more widely. Again, anomalous angular dependence of the scattering intensity and linewidth of the QLLS spectrum are observed. Contrary to the case for repulsive forces, the relative scattered intensity increases as $q \rightarrow 0$, and the effective diffusion constant $D_{\text{eff}} = \Gamma(q)/q^2$ decreases as $q \rightarrow 0$. The free-particle diffusion coefficient D_t is obtained in the small-q limit. The angular dependence of the scattered intensity $S(q)$ and the linewidth

$\Gamma(q)$ can be described in terms of a temperature-dependent correlation length for the intermolecular interaction $\xi(T)$ (*106, 107*):

$$S(q) = S(q=0)\,\frac{\xi_s(T)^{-2}}{\xi_s(T)^{-2} + q^2} \tag{78}$$

$$D(q) = D(q=0)[1 + \xi_\Gamma^2(T)q^2] \tag{79}$$

The temperature dependence of the experimental data is usually characterized in terms of critical exponents (*108*) which characterize the deviations of the system from the simple van der Waals model of Debye:

$$\lim_{q\to 0} S(q) \propto (\tau - 1)^{\gamma} \tag{80}$$

$$\lim_{q\to 0} D(q) \propto (\tau - 1)^{\gamma^*} \tag{81}$$

$$\xi_s \propto (T_r - 1)\nu_s \tag{82}$$

$$\xi_\Gamma \propto (T_r - 1)\nu_r \tag{83}$$

where $T_r = T/T_c$, T_c being the critical temperature for demixing. An extension of this critical exponent concept has been introduced by Benedek (*3*) and Chu et al. (*109*) which relates the temperature dependence of the static and dynamic properties to a pseudo-spinodal state imagined to represent the limit of stability of the metastable single-phase system:

$$\lim_{q\to 0} S(q) \propto (T_r - T_{sp})^{\gamma} \tag{84}$$

$$\lim_{q\to 0} D(q) \propto (T_r - T_{sp})^{\gamma^*} \tag{85}$$

Chu and co-workers (*110, 111*) have studied the photon correlation function of light scattered by solutions of narrow molecular weight fractions of PS (M_n = 397,000) in cyclohexane near the critical demixing temperature T_c. Their data fit Equations 82 and 83 with $\gamma = 1.26 \pm 0.08$, $\gamma^* = 0.77$. The correlation lengths were in the range of 100–250Å, and $\nu_s = 0.62$, $\nu_\Gamma = 0.58$. For $(T - T_c) < 0.1°$, deviations from the critical exponent concept were discovered which may result from macromolecular polydispersity or configurational fluctuations (*111*).

In our laboratory, we have recently carried out a study of the QLLS spectrum of light scattered by dilute aqueous solutions containing mixtures of the oppositely charged macroions poly-L-lysine HBr and chondroitin-6-sulfate (*112, 113*). Because of the electrostatic interactions, large intermacromolecular aggregates are formed, and under certain conditions of pH and ionic strength, visible turbidity and a rapid time-dependent decrease of the apparent diffusion coefficient $D_{eff} = \Gamma(q)/q^2$ is observed, indicating incipient precipitation. In all of the solutions studied, anomalous angle dependence of the scattering intensity and effective diffusion coefficient is observed, as shown in Figure 5. Preliminary analysis of this evidence according to Equations 78 and 79 lead to the conclusion that $\xi = 140$Å. Alternative interpretations are possible, however. First, angular dependence of the effective diffusion coefficient as depicted in Figure 5 would be observed if the intermacromolecular aggregates are nonspherical and exhibited rotational diffusion with $D_r =$

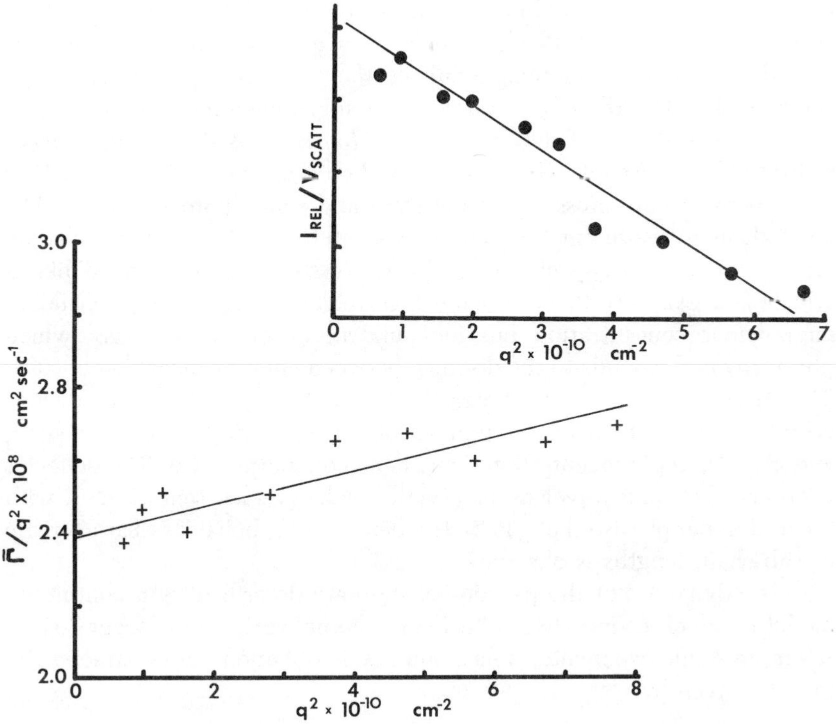

Figure 5. Effective translational diffusion coefficient $D_{eff} = \overline{\Gamma}/q^2$ *of mixtures of poly-L-lysine HBr and chondroitin 6-sulfate in aqueous 0.1M NaCl at 25°C: Lysine to disaccharide ratio is 1:1. Inset is the variation in relative scattered intensity per unit scattering volume with the square of the wave vector q for the same system.*

75 sec^{-1}. Second, polydispersity of the aggregates could be interpreted as responsible for the angular dependence of $\Gamma(q)/q^2$ of the system, as evident by comparing Figure 3 with Figure 5. This interpretation is usually excluded on the basis of the accurate Lorentzian shape of the experimental spectra (*32, 107*). We have noted elsewhere that our small angle spectra are remarkably close to single Lorentzian functions at small angles (*112, 113*), but not at angles θ beyond 60°. Because of this and the fact that the nonlinear dependence of Γ/q^2 on q^2 persists on dilution of the system ,we presently tend to interpret the phenomenon as resulting from rotational diffusion. If the latter interpretation turns out to be correct for the system represented in Figure 5, an appropriate model for the hydrodynamic structure of the aggregates would be a prolate ellipsoid with semi-axes: $a = 2250$Å, $b = 500$Å.

Semi-Dilute Solutions and Polymer Gels. In this section, we are concerned with polymer–solvent systems in which there is substantial overlap between neighboring polymer coils. Recent theoretical progress has been made in describing the static (*39*) and dynamic (*114, 115*) scattering behavior of these systems in which not only the repulsive excluded volume interactions influence the appropriate space–time correlations, but also the effects of chain entanglements must be taken into account. The theory of deGennes (*114, 115*) extends the earlier analysis of Edwards and Freed (*116*) and is based on the concept that an analogy exists between the cross-linked gel state and semi-dilute solutions. The entanglements represent tie points whose lifetime extends over a characteristic time T_r. For times $\tau < T_r$, the system should behave like a crosslinked gel. The theory defines a correlation length $\xi(c)$, dependent on polymer concentration but independent of molecular mass, which physically corresponds to the distance between entanglement points (*39*). The dynamic behavior encountered in a scattering experiment depends critically on the relative magnitudes of the scattering vector q, $\xi(c)^{-1}$, and $R(c)^{-1}$, the concentration-dependent coil radius. Gel-like behavior is observed at long wavelengths (small q-values) such that $q\xi < 1$, with the additional proviso that $q\mathrm{R} > 1$; when $q\xi > 1$, behavior characteristic of sub-chain lengths is observed.

The dynamics of the pseudo-gel state are described by a continuum model in which cooperative diffusion of the network chains occur (*117*). A simple single-exponential relaxation law is obtained with characteristic time Γ^{-1} given by:

$$\Gamma = \frac{E}{\Phi} q^2 \tag{86}$$

where E is the rigidity modulus of the gel, and Φ is a friction constant describing flow of solvent through the gel. According to deGennes arguments (*114, 115*):

$$E/\Phi = D_c = \frac{k_B T}{6\pi\eta_0\xi} \tag{87}$$

where D_c is a cooperative diffusion coefficient. Since static neutron scattering experiments indicate $\xi \sim c^{-3/4}$, Equation 87 indicates $D_c \sim c^{0.75}$. Under these conditions, QLLS experiments should find a single Lorentzian spectral component (single exponential decay), and this indeed is apparently the case for cross-linked polyacrylamide (*117*) and PS gels (*118*) as well as in concentrated solutions of PS (*119, 120, 121*). The latter experiments have additionally verified other features of deGennes theory such as the concentration dependence predicted by Equation 87. In some solvents, however, there are indications that the accuracy of the theoretical predictions is diminished (*121*).

The deGennes analysis (*114, 115*) predicts somewhat more complex scattering behavior at lower concentrations when $\xi^{-1} \sim q \sim R^{-1}$, which is the dynamic regime addressed in the work of Lee et al. (*104*) described in the previous section. A smooth crossover between the gel dynamical scheme and dynamics corresponding to individual chains is predicted at $q\xi \sim 1$. Small-angle dynamic scattering should reflect the continuum gel-like mode, wide-angle scattering should detect free-particle diffusive modes (suitably modified by intramolecular relaxation behavior since $q \sim R^{-1}$). The characteristic frequency at which this behavior occurs is described by D_c/ξ^2 which turns out to be essentially identical to the fundamental Zimm mode for the independent fluctuations of the segmental distribution between entanglements. However, there is no provision in the single-mode relaxation formalism presented by deGennes for the bimodal dynamic behavior which is reported experimentally in certain polyelectrolyte systems as is discussed above.

Other workers have adopted a different corpuscular model for quasielastic light scattering from a gel (*122, 123, 124*). The gel is treated as an assembly of identical, independent, harmonically bound particles, each in Brownian motion about a stationary mean position. The analysis of Carlson and co-workers (*123, 124*) formally includes the presence of static interference scattering components resulting from spatial structuring of the polymer chains and from the consequent constraints in the diffusive motions of the chains. This formalism leads to the prediction of nonexponential scattered intensity autocorrelation functions.

Experimental results to support both theoretical models have been reported. McAdam et al. (*122*) and Carlson et al. (*123, 124*) found nonexponential decays of the correlation function in studies of cross-linked PS gels swollen in tetralin and cross-linked aqueous polyacrylamide gels, respectively. On the other hand, the data of Tanaka et al. (*122*) for aqueous polyacrylamide gels agree with the continuum hydrodynamic theory as in later data by Munch et al. (*118*) on polystyrene gels in benzene. The reason for the discrepancies between these different experimental results is still unclear. Recently Munch et al. (*125*) showed that the correlation data for polystyrene gels with an artificially polydisperse distribution of chain lengths between branch points are still accurately single exponential decays. This would seem to indicate the source of nonexponential behavior must indeed lie in the macrostructural heterogeneity of the gels.

Acknowledgments

We gratefully acknowledge research support from the National Institutes of Health (NIAMD-17110 and HL-15195) and the National Science Foundation (ENG 76-20278 and DMR 76-82465, Polymers Program).

Literature Cited

1. Berne, B. J., Pecora, R., "Dynamic Light Scattering," Wiley, New York, 1976.
2. Chu, B., *Annu. Rev. Phys. Chem.* (1970) **21**, 145.
3. Benedek, G. B., "Polarization, Matter and Radiation," p. 49, Presses Universitaires de France, Paris, 1969.
4. Cummins, H. Z., Swinney, H. L., *Prog. Opt.* (1970) **8**, 133.
5. Pecora, R., *Annu. Rev. Biophys. Bioeng.* (1972) **1**, 257.
6. Jamieson, A. M., Maret, A. R., *Chem. Soc. Rev.* (1973) **2**, 325.
7. Ford, N. C., Gabler, R., Karasz, F. E., ADV. CHEM. SER. (1973) **125**, 25.
8. Ford, N. C., *Chem. Scr.* (1972) **2**, 193.
9. Dubin, S. B., *Methods Enzymol.* (1972) **26**, 119.
10. Ware, B. R., *Adv. Colloid Interface Sci.* (1974) **4**, 1.
11. Carlson, F. D., *Annu. Rev. Biophys. Bioeng.* (1975) **4**, 243.
12. Elson, E. L., Webb, W. W., *Annu. Rev. Biophys. Bioeng.* (1975) **4**, 311.
13. Chu, B., "Laser Light Scattering," Academic, New York, 1974.
14. "Photon Correlation and Light Beating Spectroscopy," H. Z. Cummins, E. R. Pike, Eds., Plenum, New York, 1974.
15. Casassa, E. F., Eisenberg, H., *Adv. Protein Chem.* (1964) **19**, 287.
16. Eisenberg, H., *J. Chem. Phys.* (1962) **36**, 1837.
17. Huglin, M. B., Ed., "Light Scattering from Polymer Solutions," Academic, New York, 1972.
18. Van Hove, L., *Phys. Rev.* (1954) **95**, 249.
19. Pecora, R., *J. Chem. Phys.* (1968) **49**, 1036.
20. Schaefer, D. W., Benedek, G. B., Schofield, P., Bradbury, E., *J. Chem. Phys.* (1971) **55**, 3884.
21. Aragon, S. R., Pecora, R., *J. Chem. Phys.* (1976) **64**, 2395.
22. Phillies, G. D. J., Benedek, G. B., Mazer, N. A., *J. Chem. Phys.* (1976) **65**, 1883.
23. Schaefer, D. W., Berne, B. J., *Phys. Rev. Lett.* (1972) **28**, 475.
24. Jamieson, A. M., Walton, A. G., *J. Chem. Phys.* (1973) **58**, 1054.
25. Brown, J. C., Pusey, P. N., Dietz, R., *J. Chem. Phys.* (1975) **62**, 1136.
26. McDonnell, M. E., Jamieson, A. M., *Biopolymers* (1976) **15**, 1283.
27. Pecora, R., Tagami, Y., *J. Chem. Phys.* (1969) **51**, 3298.
28. Frederick, J. E., Reed, T. F., Kramer, O., *Macromolecules* (1971) **4**, 242.
29. Lee, S. P., Chu, B., *Appl. Phys. Lett.* (1974) **24**, 261.

30. Chen, F. C., Yeh, A., Chu, B., *J. Chem. Phys.* (1976) **65**, 4509.
31. Koppel, D. E., *J. Chem. Phys.* (1972) **57**, 4814.
32. Tanaka, T., *Polym. J.* (1975) **7**, 62.
33. Brehm, G. A., Bloomfield, V. A., *Macromolecules* (1975) **8**, 663.
34. King, T. A., Knox, A., Lee, W. I., McAdam, J. D. G., *Polymer* (1973) **14**, 151.
35. Chen, F. C., Chrzeszczyk, A., Chu, B., *J. Chem. Phys.* (1976) **64**, 3403.
36. Ibid. (1977) **66**, 2237.
37. McDonnell, M. E., Jamieson, A. M., *J. Macromol. Sci., Phys.* (1977) **B13**, 67.
38. Yamakawa, H., "Modern Theory of Polymer Solutions," Chap. VI, Harper and Row, New York, 1971.
39. Daoud, M., Cotton, J. P., Farnoux, B., Jammink, G., Sarma, G., Benoit, H., Duplessix, R., Picot, C., deGennes, P. G., *Macromolecules* (1975) **8**, 804.
40. Gosting, L. J.,*Adv. Protein Chem.* (1956) **11**, 429.
41. McDonnell, M. E., Jamieson, A. M., unpublished data (1976).
42. Yokoyama, K., Suzuki, A., Utsumi, I., Naito, R., *Chem. Pharm. Bull.* (1974) **22**, 2966.
43. Ford, N. C., Karasz, F. E., Owen, J. E. M., *Discuss. Faraday Soc.* (1973) **49**, 228.
44. Mandelma, W., Zeldenrust, H., *Polymer* (1977) **18**, 835.
45. Castellino, F. J., Barker, R., *Biochemistry* (1968) **7**, 2207.
46. Dubin, S. B., Benedek, G. B., Bancroft, F. C., Frifelder, D. J., *J. Mol. Biol.* (1970) **54**, 547.
47. Camerini-Otero, R. D., Pusey, P. N., Koppel, D. E., Schaefer, D. W., Franklin, R. M., *Biochemistry* (1974) **13**, 960.
48. Newman, J., Swinney, H. L., Berkowitz, S. A., Day, L. A., *Biochemistry* (1974) **13**, 4832, 4835.
49. Koppel, D. E., *Biochemistry* (1974) **13**, 2712.
50. Lowenstein, M. A., Birnboim, M. H., *Biopolymers* (1974) **14**, 419.
51. Lim, T. K., Baran, G. J., Bloomfield, V. A., *Biopolymers* (1977) **16**, 1473.
52. Price, C., McAdam, J. D. G., Lally, T. P., Woods, D., *Polymer* (1974) **15**, 228.
53. Spach, G., Freund, L., Daune, M., Benoit, H., *J. Mol. Biol.* (1963) **7**, 468.
54. Perrin, F., *J. Phys. Radium* (1934) **5**, 497.
55. Benoit, H., Freund, L., Spach, G., "Poly-α-Amino Acids," G. D. Fasman, Ed., p. 105, Arnold, London, 1967.
56. Wright, A. K., Baxter, J. E., *Biophys. J.* (1976) **16**, 931.
57. Dubin, S. B., Clark, N. A., Benedek, G. B., *J. Chem. Phys.* (1971) **54**, 5158.
58. Ford, N. C., Lee, W., Karasz, F. E., *J. Chem. Phys.* (1969) **50**, 3098.
59. Jamieson, A. M., Mack, L., Walton, A. G., *Biopolymers* (1972) **11**, 2267.
60. Lee, W. I., Schurr, J. M., *Biopolymers* (1974) **13**, 903.
61. Patkowski, A., Chu, B., "Abstracts of Papers," 173rd National Meeting, ACS, March 21–25, 1977, COLL 079.
62. Kirkwood, J. G., Riseman, J., *J. Chem. Phys.* (1948) **16**, 565.
63. Stockmayer, W. H., Albrecht, A. C., *J. Polym. Sci.* (1958) **32**, 215.
64. Fair, B. D., Chao, D.-Y., Jamieson, A. M., *J. Colloid Interface Sci.* (1978) **66**, 323.
65. McCammon, J. A., Deutch, J. M., Bloomfield, V. A., *Biopolymers* (1975) **14**, 2479.
66. Cummins, H. Z., Carlson, F. D., Herbert, T. J., Woods, G., *Biophys. J.* (1969) **9**, 518.
67. Wada, A., Suda, N., Tsuda, T., Soda, K., *J. Chem. Phys.* (1969) **50**, 31.
68. Wada, A., *Biopolymers* (1974) **13**, 237.
69. Schurr, J. M., Schmitz, K. S., *Biopolymers* (1973) **12**, 1021.

70. Pecora, R., *J. Chem. Phys.* (1968) **49**, 1032.
71. Zimm, B., *J. Chem. Phys.* (1956) **24**, 269.
72. Huang, W.-N., Frederick, J. E., *Macromolecules* (1974) **7**, 34.
73. Schmitz, K. S., Schurr, J. M., *Biopolymers* (1973) **12**, 1543.
74. Schurr, J. M., *Biopolymers* (1977) **16**, 461.
75. Pyun, C. W., Fixman, M., *J. Chem. Phys.* (1964) **41**, 937.
76. Phillies, G. D. J., *J. Chem. Phys.* (1974) **60**, 976.
77. Ibid. (1974) **60**, 983.
78. Batchelor, G. K., *J. Fluid Mech.* (1976) **74**, 1.
79. Anderson, J. L., Reed, C. C., *J. Chem. Phys.* (1976) **64**, 3240.
80. Alpert, S. S., *J. Chem. Phys.* (1976) **65**, 4333.
81. Keller, K. H., Canales, E. R., Yum, S. I., *J. Phys. Chem.* (1971) **75**, 379.
82. Tanford, C., Swanson, S. A., Shore, W. S., *J. Am. Chem. Soc.* (1955) **77**, 6414.
83. Imai, S., *J. Chem. Phys.* (1970) **53**, 4212.
84. Freed, K. F., *J. Chem. Phys.* (1976) **65**, 4103.
85. Lacharojana, S., Caroline, D., *Macromolecules* (1977) **10**, 365.
86. Frost, R. A., Caroline, D., *Macromolecules* (1977) **10**, 616.
87. Vrentas, J. S., Duda, J. L., *J. Polym. Sci., Polym. Phys. Ed.* (1976) **14**, 101.
88. Noda, I., Mizutani, K., Kato, T., *Macromolecules* (1977) **10**, 618.
89. Herbert, T. J., Carlson, F. D., *Biopolymers* (1971) **10**, 2231.
90. Chu, B., Yeh, A., Chen, F. C., Weiner, B., *Biopolymers* (1975) **14**, 93.
91. Jamieson, A. M., Presley, C. T., *Macromolecules* (1973) **6**, 358.
92. Pusey, P. N., Schaefer, D. W., Koppel, D. E., Camerini-Otero, R. D., Franklin, R. M., *J. Phys. (Paris) Colloq.* (1972) **33**, C1.
93. Schaefer, D. W., Berne, B. J., *Phys. Rev. Lett.* (1974) **32**, 1110.
94. Brown, J. C., Pusey, P. N., Goodwin, J. W., Ottewill, R. H., *J. Phys. A* (1975) **8**, 664.
95. Schmidt, R. L., *Biopolymers* (1973) **12**, 1427.
96. Pusey, P. N., *J. Phys. A* (1975) **8**, 1433.
97. Ackerson, B. J., *J. Chem. Phys.* (1976) **64**, 242.
98. Zwanzig, R., *Adv. Chem. Phys.* (1969) **15**, 325.
99. Schneider, T., *Phys. Rev.* (1971) **A3**, 2145.
100. deGennes, P. G., *Physica* (1959) **25**, 825.
101. Schmitz, K. S., Pecora, R., *Biopolymers* (1975) **14**, 521.
102. Schurr, J. M., *Q. Rev. Biophys.* (1976) **9**, 109.
103. Lin, S. C., Schurr, J. M., Schmitz, K. S., *Biophys. J.* (1976) **16**, 59a.
104. Lee, W. I., Schmitz, K. S., Lin, S.-C., Schurr, J. M., *Biopolymers* (1977) **16**, 583.
105. Solc, K., *J. Chem. Phys.* (1971) **55**, 335.
106. Debye, P., *J. Chem. Phys.* (1959) **31**, 680.
107. Chu, B., Schoenes, F. J., *Phys. Rev. Lett.* (1968) **21**, 6.
108. Kadanoff, L. P., Gotze, W., Hamblen, D., Hecht, R., Lewis, E. A. S., Palcianskas, V. V., Raye, M., Swift, J., *Rev. Mod. Phys.* (1967) **39**, 395.
109. Chu, B., Schoenes, F. J., Fisher, M. E., *Phys. Rev.* (1969) **185**, 219.
110. Kuwahara, N., Fenby, D. V., Tamsky, M., Chu, B., *J. Chem. Phys.* (1971) **55**, 1140.
111. Lee, S. P., Tscharnuter, W., Chu, B., Kuwahara, N., *J. Chem. Phys.* (1972) **57**, 4240.
112. Schodt, K. P., McDonnell, M. E., Jamieson, A. M., Blackwell, J., *Macromolecules* (1977) **10**, 701.
113. Schodt, K. P., Ph.D. Thesis, Case Western Reserve University (1977).
114. deGennes, P. G., *Macromolecules* (1976) **9**, 587.
115. Ibid., 594.
116. Freed, K., Edwards, S. F., *J. Chem. Phys.* (1974) **61**, 3626.
117. Tanaka, T., Hocker, L. O., Benedek, G. B., *J. Chem. Phys.* (1973) **59**, 5151.

118. Munch, J. P., Candau, S., Duplessix, R., Picot, C., Herz, H., Benoit, H., *J. Polym. Sci., Polym. Phys. Ed.* (1976) **74,** 1097.
119. Geissler, E., Hecht, A. M., *J. Chem. Phys.* (1976) **65,** 103.
120. Ibid. (1976) **66,** 1416.
121. Adam, M., Delsanti, M., Jannink, G., *J. Phys. (Paris)* (1976) **37,** L-53.
122. McAdam, J. D. G., King, T. A., Knox, A., *Chem. Phys. Lett.* (1974) **26,** 6.
123. Carlson, F. D., Fraser, A., *J. Mol. Biol.* (1974) **89,** 273.
124. Wun, K. L., Carlson, F. D., *Macromolecules* (1975) **8,** 190.
125. Munch, J. P., Candau, S., Hild, G., *J. Polym. Sci., Polym. Phys. Ed.* (1977) **15,** 11.

RECEIVED November 10, 1977.

11

Studies of Polymer Dynamics by Multipass Fabry–Perot Spectroscopy

S. M. LINDSAY[1] and I. W. SHEPHERD

Physics Department, University of Manchester,
Manchester, M13 9PL, United Kingdom

Brillouin scattering has been used to observe several hypersonic relaxations in polydimethyl siloxane (PDMS), polyisobutylene (PIB), and polypropylene oxide (PPO), and the cross-link dependence of hypersonic relaxation in PDMS has been investigated. Activation energies for the major relaxations determined by varying scattering angles are: PDMS 11 K cal/mol (240 K relaxation), PIB 22 K cal/mol (430 K relaxation), PPO 10 K cal/mol (300 K relaxation). Reduced PPO data is fitted to a two-relaxation time model with $\tau_1 = 0.09\ \tau_M$, $\tau_2 = 0.002\ \tau_M$ *(*τ_M *is the Maxwellian relaxation time). Hypersonic loss in the networks shows a maximum at 84 monomer units per cross-link (m.u./C) at 5GHz and 35 m.u./C at 2.5 GHz. Using the Adams–Gibbs model of polymer motion the size of cooperative regions are estimated from measured activation energies and the network data qualitatively accounted for.*

The multipassed Fabry–Perot spectrometer (*1*) has allowed Brillouin scattering to be studied in many materials for the first time, and much of this new activity has concentrated on polymers (*see* for example the review by Patterson (*2*)). The long-wavelength acoustic phonons studied in a Brillouin experiment (where phonon wavelengths range from $\sim 0.1\,\mu$ to $\sim 10\,\mu$) provide a sensitive probe of structure on a macromolecular scale. At temperatures well above the glass transition temperature, T_g, relaxation rates become comparable with phonon frequencies

[1] Present address: Mullard Hazel Grove, Bramhall Moor Lane, Hazel Grove, Cheshire, United Kingdom.

0-8412-0406-3/79/33-174-207$05.00/1

in a Brillouin experiment (10^8–10^{10} Hz) and the technique becomes a useful probe of polymer dynamics at these temperatures.

Perhaps of the most interest is the rubber–glass transition itself and many early Brillouin studies measured the speed of sound and phonon damping around T_g (*3, 4, 5, 6*) while others attempted to observe changes in the ratio of elastically to inelastically scattered light intensities over the transition (*6, 7, 8, 9*). These measurements yielded little new information, for phonon attenuation does not change measurably over the transition while reported discontinuities in intensity ratios appear to be experimental artifacts (*2, 8*). The sound speed–temperature gradient does change over the transition, but the observed changes may be accounted for by the differences in volume expansion and Grüneisen parameters between the rubbery and glassy states (*10*). The null results of the measurements of intensity ratios and phonon damping may in themselves be significant, and we discuss this point in the final section of this chapter. From a dynamical viewpoint the important feature is that polymer rearrangements occur slowly near T_g. To study the glass transition by Brillouin scattering it is necessary to go to higher temperatures where relaxation times approach 10^{-8}–10^{-10} sec. High-temperature hypersonic loss processes have been reported in several polymers (*11–16*). However, with the exception of the loss process in polymethylacrylate (PMA) (*14*) investigation has been limited to correlating the hypersonic loss process with the main chain relaxation (α and β processes appear to merge at these frequencies) via a temperature–frequency transition map (*11, 15, 16*). In this chapter we examine the hypersonic loss in polymers, particularly polydimethyl siloxane (PDMS), polyisobutylene (PIB), and polypropylene oxide (PPO).

Several features are demonstrated:

(a) In the polymers studied there is evidence of at least two relaxations associated with the hypersonic loss. This is most clearly demonstrated by using the technique of reduced variables in analyzing Brillouin data.

(b) Activation energies for the relaxations may be directly obtained from Brillouin data taken at different scattering angles.

(c) Cross-linking a gum appears to increase the relaxation times associated with the hypersonic loss.

In discussing these results we shall attempt to relate the high-temperature–high-frequency transition to its quasi-static counterpart through the cooperative model of Adam and Gibbs (*17*).

Theory

In the Brillouin scattering experiment a photon is inelastically scattered from a phonon giving rise to two peaks in the scattered light spectrum shifted from the elastically scattered light by an amount $\pm \nu_B$. This shift is related to the speed of hypersound, V_s, by:

$$V_s = \frac{\nu_B \lambda_o}{2n \operatorname{Sin} \Theta/2} \tag{1}$$

where λ_o is the incident laser wavelength in vacuo, n is the refractive index of the sample, and Θ is the internal scattering angle. A number of relationships between acoustic loss and Brillouin linewidth appear in the literature (*5, 12, 15*), so for clarity we shall state the relationships we have used in more detail than is usual. Damping of the phonons gives rise to lines which are approximately Lorentzian in shape with a half-width at half maximum, $\nu_{\frac{1}{2}}$, equal to the temporal decay coefficient, α, for (*18*) amplitude fluctuations decaying as:

$$A(t) = A_o e^{-\Gamma t} \tag{2}$$

In a conventional sound attenuation experiment the linear intensity decay coefficient, α, is measured and clearly it is related to Γ by:

$$\alpha = \frac{2\Gamma}{V_s} \tag{3}$$

The loss per cycle, $\alpha\lambda_s$ (λ_s is the sound wavelength) is obtained from:

$$\alpha\lambda_s = \frac{2\nu_{\frac{1}{2}}}{\nu_B} \tag{4}$$

This quantity is, of course, identical to the ratio of the real and imaginary components of the complex modulus, the loss tangent. In the absence of a viscoelastic loss process $\alpha\lambda_s$ is only weakly temperature dependent having a value of approximately 0.1–0.05 arising from structural (*5*) or phonon–phonon scattering (*12, 19*). Viscoelastic relaxation gives rise to a large increase in $\alpha\lambda_s$ at the loss temperature. In cases of large $\alpha\lambda_s$ Montrose et al. (*20*) have shown that a better approximation is,

$$\alpha\lambda_s = \frac{2\Gamma^+}{\nu_B}\left[1 + \frac{\Gamma^+}{2\nu_B}\right]$$

where Γ^+ is the half-width at half-weight measured on the high-frequency side of the peak.

Experimental

Spectra were recorded on a multipass Fabry–Perot spectrometer, described in detail elsewhere (*11, 12, 22*), using the 514.5 nM line of an Ar^+ laser. The spectrometer was operated at an optimized free spectral range in each case with a finesse of approximately 40. Spectra were recorded digitally and instrumental distortions were corrected by the method of Lindsay, Burgess, and Shepherd (*23*). Temperatures above 293 K were obtained with an electrically heated jacket around the sample cell and the temperature was measured with a thermocouple placed in the sample near the scattering volume. Lower temperatures were obtained with an Oxford Instruments CF-104 light-scattering cryostat. Temperature control and measurement was accurate to 0.1°C.

The samples used were ICI S273 11–3 (PDMS), Eso Vistanex LH–MH (PIB), and Shell PPG 1500 (PPO). The weight average molecular weights measured by gel permeation chromatography (GPC) were 77,000, 70,000, and 1,600, respectively. Rubber networks were formed by radiation cross-linking as described elsewhere (*24*) giving networks with shear moduli up to approximately 10^6 N/M^2 as measured by ball-bearing indentation (*25*). Refractive indices were measured on an Abbé refractometer and at 20°C are 1.403 (PDMS), 1.501 (PIB), 1.449 (PPO). Their temperature variation was small (less than 5% over the range studied) but it was taken into account in calculating V_s. Values of $\alpha\lambda_s$ were obtained to ± 7% accuracy.

Experimental Results and Analysis

Velocity and Attenuation Data from Polymer Fluids. PDMS. V_s and $\alpha\lambda_s$ were measured over the range 220–360 K at internal scattering angles of 74° and 158° so that phonon frequencies varied from 4.2 to 3 GHz and 6.85 to 4.9 GHz, respectively. (Cryostat geometry limited the available range of scattering angles.) Values of $\alpha\lambda_s$ are plotted in Figure 1a and V_s in Figure 1b where only the 74° velocity data are shown for clarity. (The 158° velocity data lie on the same curve to within measurement error.) Loss maxima are observed at about 235 K, 250 K, 275 K, and 320 K in the low-angle data while the sound-speed data indicate a major relaxation at about 240 K with some evidence for a minor relaxation at approximately 330 K. In the high-angle loss data the largest peak is shifted to about 240 K, and the 275 K peak (at 4 GHz) appears to have shifted to about 290 K. The others, though broadened, appear unshifted.

PIB. Values of $\alpha\lambda_s$ and V_s are plotted in Figures 2a and 2b over the temperature range 300–520 K for internal scattering angles of 125° and 20°. At the lower angle the F16 collection optics contributed about 4% to the linewidth in the worst case, and our data has been corrected for finite aperture broadening (*26*). Phonon frequencies varied from 13.5 to 6.3 and 2.7 to 1.2 GHz in the high- and low-angle experiments, respectively. High-angle loss data show a maximum at about 460 K with some

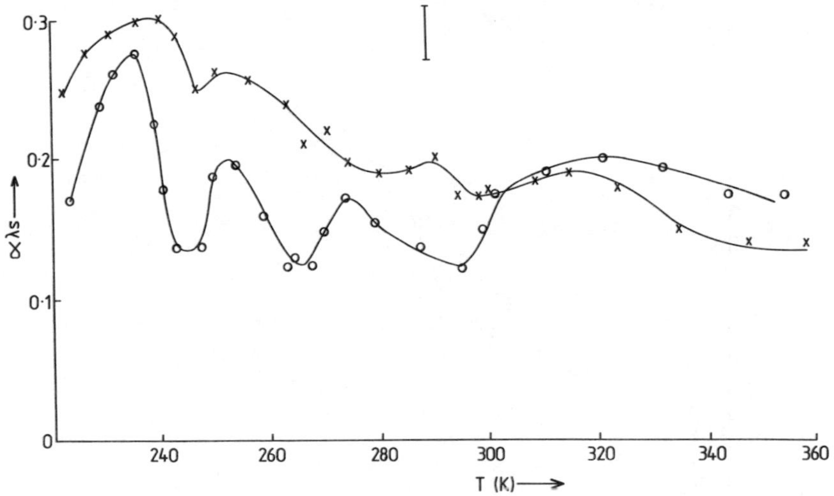

Figure 1a. Loss per cycle $\alpha\lambda_s$, in PDMS fluid as a function of temperature: (○), 74° scattering angle; (×), 158° scattering angle. For clarity only one representative error bar is shown on this and subsequent figures.

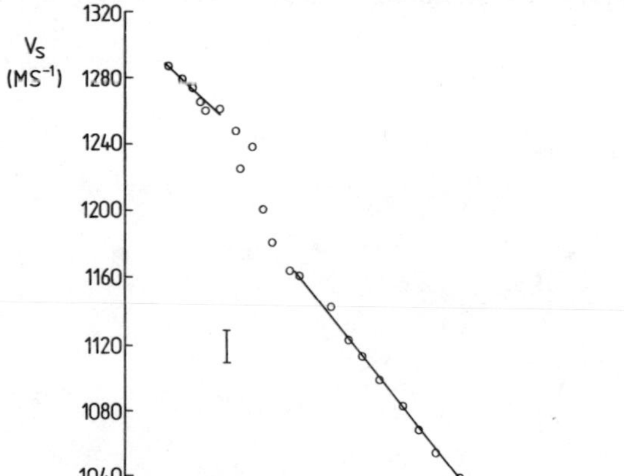

Figure 1b. Sound speed in PDMS fluid as a function of temperature: (○), 74° scattering angle; (×), 158° scattering angle. For clarity only one representative error bar is shown on this and subsequent figures.

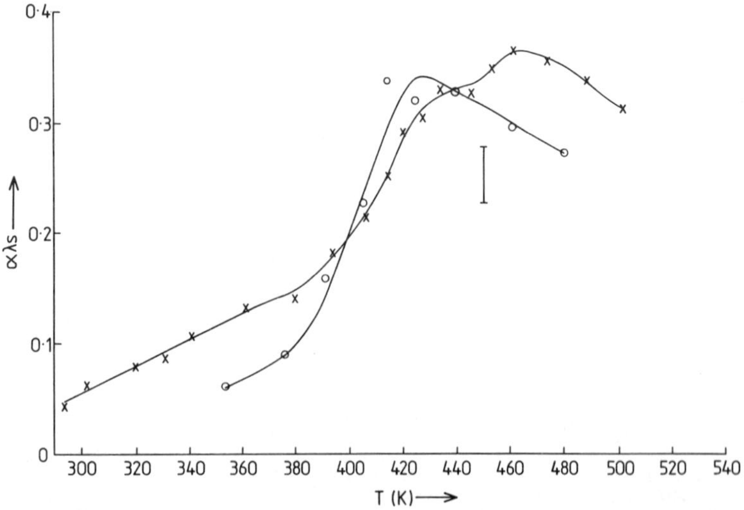

Figure 2a. Loss per cycle, $\alpha\lambda_s$, in PIB fluid as a function of temperature: (○), 20° scattering angle; (×), 125° scattering angle

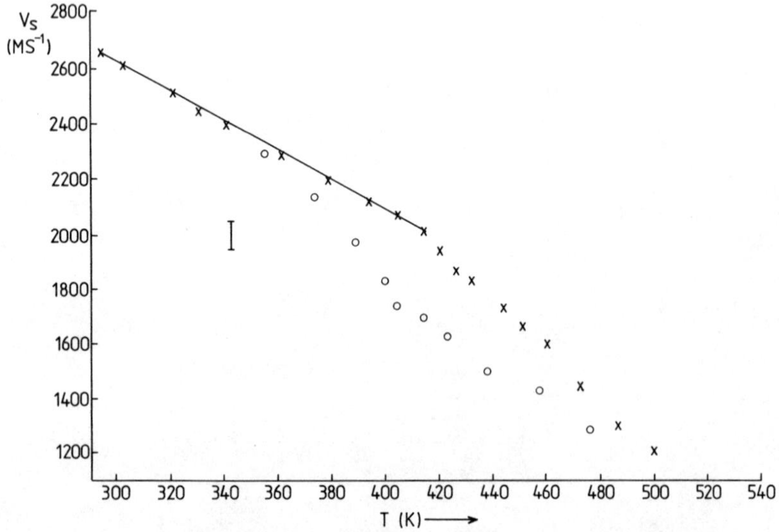

Figure 2b. Sound speed in PIB fluid as a function of temperature: (○), 20° scattering angle; (×), 125° scattering angle

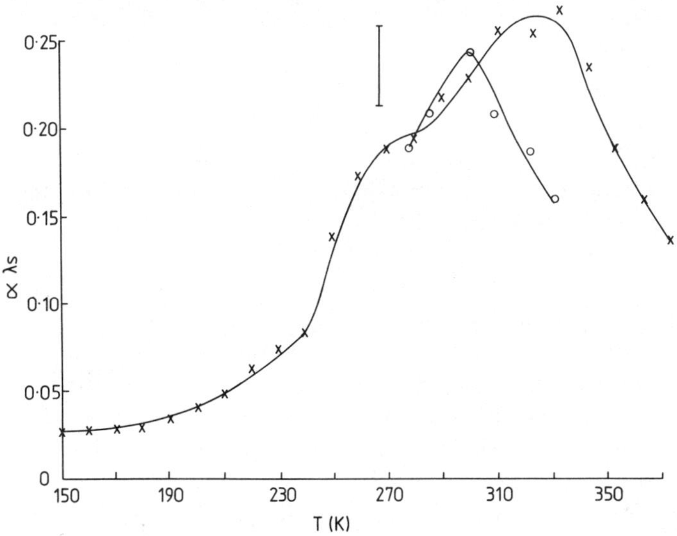

Figure 3a. Loss per cycle, $\alpha\lambda_s$, in PPO fluid as a function of temperature: (○), 20° scattering angle; (×), 59° scattering angle

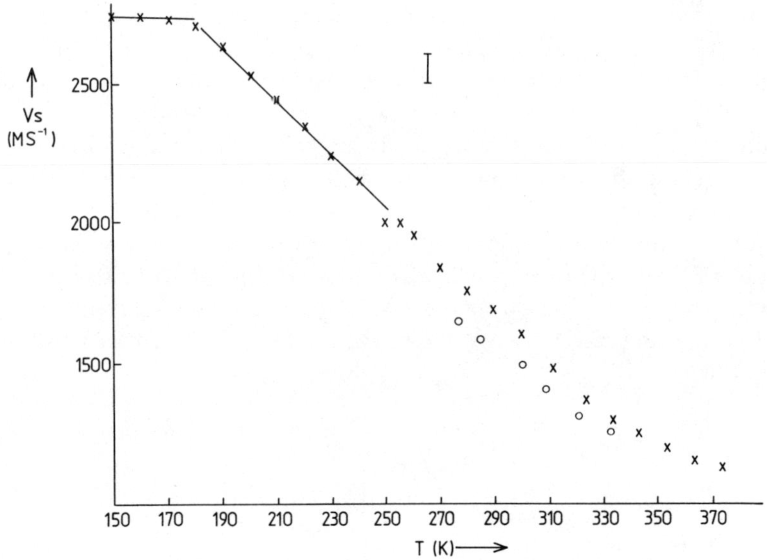

Figure 3b. Sound speed in PPO fluid as a function of temperature: (○), 20° scattering angle; (×), 59° scattering angle

evidence of a subsidiary maximum at approximately 430 K. In the low-friequency experiment our data is sufficient only to locate one maximum at about 430 K.

PPO. Measurements taken at an internal scattering angle of 59° (Figure 3a) show a loss maximum at about 320 K with a subsidiary maximum at about 270 K. Data taken at a 20° scattering angle were limited to temperatures above 273 K, none the less the loss maximum is clearly visible at approximately 300 K.

Phonon frequencies corresponding to the velocity data in Figure 3b were 7.5–3 GHz in the high-angle measurements and 1.5–1 GHz in the low-angle measurements. The high-angle data encompass the static glass transition at about 180 K, which is clearly seen as a break in the velocity temperature gradient. Before the onset of relaxation this gradient is constant showing solid-like behavior well above T_g. Near T_g the slope can be fitted to a solid-like model assuming the reasonable value of 4.5 for the Grüneisen parameter (*10*).

Multiple Relaxations. In the case of PDMS multiple relaxation is fairly clear despite measurement uncertainties. This is less evident in the cases of PIB and PPO. A clearer presentation of the hypersonic relaxation may be obtained through the method of reduced variables (*27*). In Brillouin-scattering experiments this method is essential, for unlike an ultrasonic experiment, the measurement frequency may vary by a factor of more than two over the temperature range of interest. Short of changing the scattering angle at each temperature the distortion owing to the changing Brillouin shift can easily be corrected only by means of a reduced plot. In reducing our data we have followed the method of Pinnow et al. (*28*) who first applied this technique to a Brillouin study of glycerol. Reduction requires knowledge of the temperature dependence of the limiting value of one of the moduli, of the density, and of the viscosity. This information is available for a series of low-molecular-weight PPO's in the ultrasonic shear measurements of Barlow et al. (*29*). Interpolating their data on the basis of number average molecular weight plotted on a logarithmic scale we arrive at the following relationship for PPG 1500:

$$J_\infty = 1.1 \times 10^{-10} + 0.16 \times 10^{-10}(T - 176) \quad (6)$$

and

$$\ln \eta_s = -8.4 + 890/(T - 176) \quad (7)$$

where J_∞ is the infinite frequency shear compliance in $N^{-1}m^2$, η_s the shear viscosity in Pascal seconds, and T is the temperature in Kelvin. We find the density to vary as:

$$\rho(T) = 1.08 \times 10^3(1 - 0.0007(T - 175)\ \mathrm{Kgm^{-3}} \qquad (8)$$

The nonrelaxaing component of the longitudinal storage modulus is simply the static bulk modulus K_o ($K_o = M_o$ since static shear cannot be sustained in a fluid), and Pinnow et al. (*28*) subtracted this quantity from their measured value of storage modulus in order to normalize the data (i.e., ensure the relaxational modulus goes to zero at zero frequency/infinite temperature). Choosing a value for K_o from the Brillouin data is not straight forward. Although the velocity temperature gradient reaches a constant value at higher temperatures (*30*) the observed modulus still appears to depend weakly on temperature through the dependence of mode frequency on specific volume (the Grüneisen relationship). We have therefore chosen the somewhat arbitrary expedient of using the sound speed, V_o, measured at the highest temperature point to define a value for K_o through:

$$\rho V_o^2 = K_o \qquad (9)$$

which gives a value of $1.2 \times 10^9\ \mathrm{N\ m^{-2}}$. Consequently, the absolute values of reduced moduli are subject to some uncertainty. However, as long as the basic assumption that the ratio of bulk-to-shear components is temperature independent holds (*27*) the profile of the relaxation is unaffected by this choice for K_o We obtain our reduced modulus plots by plotting $(M' - K_o)J_\infty$ and $M''J_\infty$ with $\omega\eta_s J_\infty$ ($\eta_s J_\infty$ is the Maxwellian relaxation time, τ_M). M' and M'' are the real and imaginary longitudinal moduli obtained from (*31*):

$$M' = \rho V^2 \qquad (10)$$

and

$$M'' = \rho V^2 \alpha \lambda_s \qquad (11)$$

The ω is the Brillouin shift in radians per second. The reduced real and imaginary modulus data are shown in Figures 4a and 4b. The plateau in $(M' - K)J_\infty$ extends to $\omega\eta_s J_\infty \sim 10^5$. In the region of solid-like behavior reduction fails below 220 K. However, the juxtaposition of the low- and high-angle reduced data and the extent of the plateau region over some three decades in $\omega\eta_s J_\infty$ lends support to the reduction procedure used. The solid curve is a single-relaxation model with a lumped relaxation time, $\bar{\tau}$, of $\bar{\tau} = 0.06\ \tau_M$. The dashed curve is based on a two-relaxation-time model with $\tau_1 = 0.09\ \tau_M$ and $\tau_2 = 0.002\ \tau_M$.

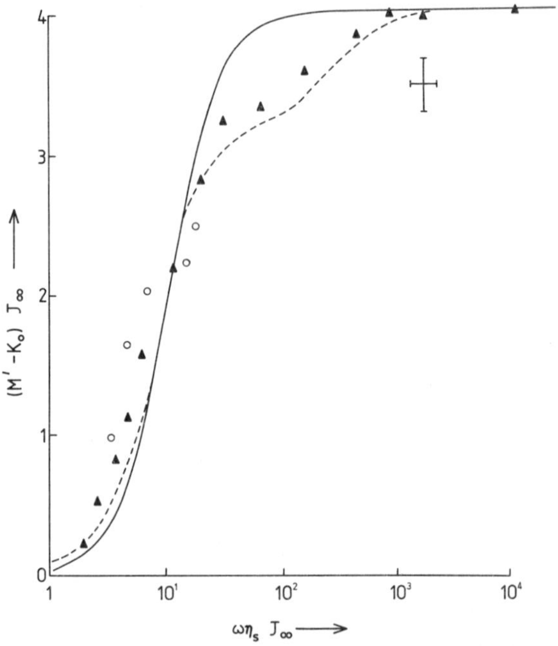

Figure 4a. Real reduced modulus plotted with the Maxwellian relaxation time–frequency product, $\omega\eta_s J_\infty$ for PPO: (▲), 59° scattering angle; (○), 20° scattering angle; (——), a best fit single relaxation; (– – –), a two-relaxation time model with $\tau_1 = 0.09\ \tau_M$, $\tau_2 = 0.002\ \tau_M$

While a broader distribution of relaxation times would provide a better fit to the experimental data, a reasonable description of the hypersonic loss process is of a major relaxation with τ approximately equal to 2.5×10^{10} sec and a secondary process with $\tau = 5 \times 10^{12}$ sec at 320 K.

Insufficient ultrasonic data are available for an analysis of the relaxation in PIB, nonetheless the raw data (Figures 2a and 2b) indicate a similar process. The behavior of PDMS is more complicated. However, at least two relaxations appear to be associated with the major hypersonic loss process.

Activation Energies. It has been shown elsewhere (*11, 15, 16*) that the major hypersonic relaxation may be located on a temperature–frequency transition map as the high-temperature glass transition, and it is interesting to obtain effective activation energies for this process. This assignment of the major loss peak assumes the α and β processes to be merged at these frequencies and recent work by Patterson (*30*) indicates that in PPO this may not be so. The uncertainty in the assignment of the major loss peak is discussed in a later section. However, we present here

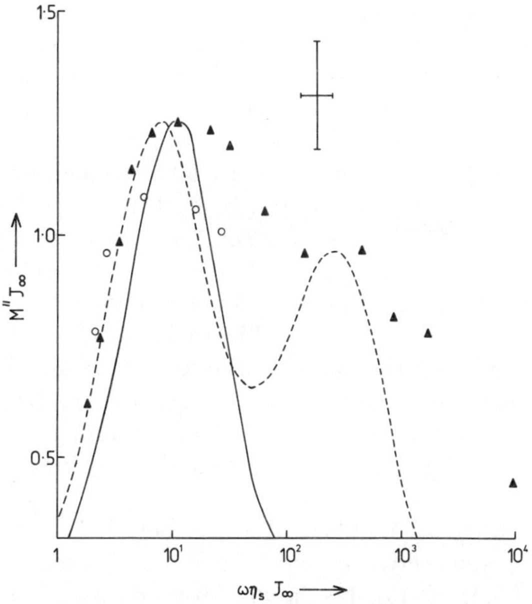

Figure 4b. Imaginary reduced modulus plotted with the Maxwellian relaxation time–frequency product, $\omega\eta_s J_\infty$ for PPO: (▲), 59° scattering angle; (○), 20° scattering angle; (——), a best fit single relaxation; (– – –), a two-relaxation time model with τ_1 0.009 τ_M, $\tau_2 = 0.002\ \tau_M$

our measurements for the strongest loss peak we observe. In the case of PDMS a measurement of the activation energy, ΔH, for the loss process is a valuable aid to assignment of the relaxation because of the dearth of high-frequency data on this polymer (*16*).

Assuming a simple Arrhenius relationship of the form,

$$\omega = \omega_0 e^{-\frac{\Delta H}{RT}} \tag{12}$$

where ω is the frequency at the loss maximum we tabulate below values of ΔH estimated from the data of Figures 1a, 2a, and 3a for the major relaxation

	ω_1 *(Hz)*	T_1 *(K)*	ω_2 *(Hz)*	T_2 *(K)*	*ΔH(K cal/ mol)*
PDMS	4.1×10^9	235 ± 3	6.7×10^9	240 ± 3	11
PIB	1.6×10^9	430 ± 5	8.3×10^9	460 ± 5	22
PPO	1.4×10^9	300 ± 5	4.0×10^9	320 ± 5	10

Small errors in temperature give rise to large uncertainties on the high side of the calculated activation energy—in the case of PDMS only a lower limit can be set on ΔH with certainty although the value given is probably of the right order. The possible range of values is: PDMS $+\infty, -5$; PIB $+10, -6$; PPO $+10, -4$ K cal/mol.

High-frequency data accumulated by other techniques indicate activation energies in this temperature range of about 10 K cal/mol (PDMS) (*16*), approximately 15 K cal/mol (PIB) (*32*), and about 20 K cal/mol (PPO) (*33*), which is in reasonable agreement with our estimates. Our data do not permit an estimate of ΔH for the minor relaxations, with the exception of the 275 K peak in the PDMS low-angle data (Figure 1a) which would appear to have an activation energy of about 5 K cal/mol. It may be associated with methyl group rotation which has an activation energy of 5.9 K cal/mol in polymethyl methacrylate (PMMA) (*34*). Studies of substituted phenyl siloxanes are under way in order to elucidate this point.

The Effect of Cross-Linking. We give here only a brief resume of our measurements on networks which have been presented elsewhere in greater detail (*35*). In Figures 5a and 5b sound speed and attenuation are plotted with equilibrium shear modulus for phonons of frequencies of about 2.5 and 5 GHz corresponding to internal scattering angles of 52° and 123°. All measurements were taken at 293 K. Chain lengths between cross-links varied from 680 to 25 monomer units (m.u.) as calculated from the measured shear modules. In highly cross-linked rubbers, values calculated from the radiation dose used in cross-linking (*36*) are in poor agreement with these values. We believe the cross-link density derived from the shear modulus to be more reliable (*24*).

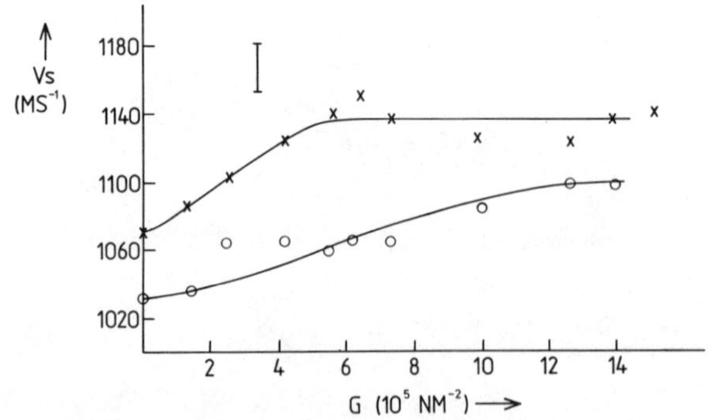

Figure 5a. Sound speed in PDMS networks plotted as a function of shear modulus, G: (○), 52° scattering angle; (×), 123° scattering angle. All data is at room temperature (293 K).

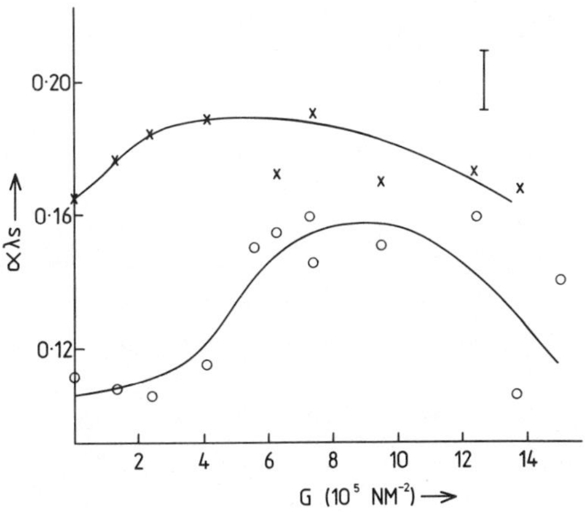

Figure 5b. Loss per cycle, $\alpha\lambda_s$, in PDMS networks plotted as a function of shear modulus, G: (○), 52° scattering angle; (×), 123° scattering angle. All data is at room temperature (293 K).

The velocity data show evidence of a strong mechanical relaxation at moderate cross-link density in the high-frequency measurements which may move towards a higher cross-link density at the lower frequency. (The increase in velocity cannot be explained by the increase in shear modulus—this represents only about 1% increase in the longitudinal modulus at the highest cross-link densities.) This movement is confirmed by absorption data (Figure 5b) which show the relaxation to be centered at G approximately equal to 4×10^5 N m^{-2} (corresponding to about 84 m.u. between cross-links) at 5 GHz and G is approximately equal to 9×10^5 N m^{-2} (corresponding to about 35 m.u. between cross-links) at 2.5 GHz. We have ruled out any effect owing to static imperfections or an additional relaxation introduced by cross-linking the network. The former on the basis of the optical clarity of the networks and the temperature dependence of the relaxation, the latter on the basis of the frequency–chain length and temperature dependence of the relaxation (*35*).

The strong temperature dependence of the relaxation is shown in Figure 6 where V_s is plotted with temperature for a rubber with G approximately equal to 7×10^9 N m^{-2} measured at about 5 GHz. As the rubber is heated above 293 K, V_s falls rapidly to a value similar to that in the fluid (Figure 1b) at these temperatures. It is tempting to draw an analogy with the cross-link dependence of the low-frequency glass transition (*37*) and we are currently extending measurements on networks to lower temperatures to see if the α relaxation is indeed affected. Unfortunately the method of sample preparation currently used for cryostalic

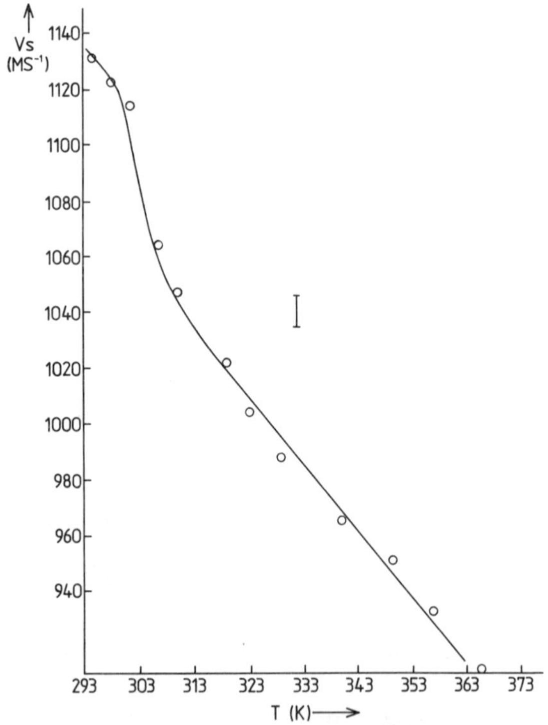

Figure 6. Sound speed in a PDMS network ($G = 7 \times 10^5$ N m^{-2}) as a function of temperature. The internal scattering angle was 160°.

work induces a degree of irreversible crystallization which adds a large background to measured values of $\alpha\lambda_s$. Differential scanning calorimetry measurements were undertaken to observe the effect of cross-linking on the low-frequency transition. However, the small shift of 4° observed in T_g in the most highly cross-linked sample is unlikely to reflect the behavior of the amorphous material since below 220 K PDMS becomes highly crystalline. These preliminary results are discussed more fully in the next section of this chapter.

Discussion

In order to achieve a coherent discussion of these results it is instructive to consider them against the background of a molecular model of polymer motion, and to do this we have chosen the cooperative model of Adam and Gibbs (*17*). We shall consider in turn the measured values of activation energy, the effects of cross-linking, and finally multiple relaxations and the assignment of loss peaks. However, before doing so it is useful to consider the results of the Brillouin measurements around T_g referred to in our introduction.

Brillouin Measurements around T_g and the Cooperative Model. Perhaps the most satisfactory model of the glass transition from both the dynamical and thermodynamic viewpoints is that of cooperatively rearranging regions whose size depends on temperature in the manner proposed by Adam and Gibbs (*17*). The number of molecules undergoing rearrangement, essentially simultaneously, Z*, is inversely related to the macroscopic entropy of the sample. The size of the cooperative region becomes macroscopic at the true thermodynamic transition temperature T_2 which is some 50° below the transition observd at T_g in a quasistatic measurement. At Brillouin frequencies the cooperative regions will appear static near T_g. However, as correlation lengths approach 0.1 μ changes in the intensity of elastically scattered light and increased phonon damping owing to structural scattering might be expected. Such effects would only be seen if the density—and hence the packing—in the cooperative regions differs from that in the amorphous phase. The fact that changes in intensity ratios or phonon damping is not seen near T_g suggests to us the important conclusion that packing in the frozen phase is indeed amorphous. The only measurements of hypersonic loss in a polymer extended to well below T_g show a gradual fall off in phonon damping right down to 4°K (*38*).

To observe the rubber–glass transition directly it is necessary to go to higher temperatures and in principle the transition can be followed upwards in temperature and frequency using a transition map. Such maps are available for many polymers, and such high-frequency information as is available indicates that the hypersonic loss peaks fall in the region of the merged α and β line (*15*).

Activation Energies. We can relate measured activation energies to the size of the cooperatively rearranging regions at the temperature of maximum loss using the following relation for relaxation rates (*17*).

$$\overline{W}(T) = \overline{A} \exp\ (Z^*\Delta\mu/kT) \tag{13}$$

$\overline{W}(T)$ is the rate of rearrangement at temperature T, A is a constant essentially independent of temperature, and $\Delta\mu$ is the barrier to rearrangement for a single monomer. A comparison with Equation 12 yields $Z^* = \Delta H/\Delta\mu$. An estimate of Z* from Equation 13 depends on the value of $\Delta\mu$ chosen. This, in turn, depends on the detailed nature of the rearrangement. However, in the case of polymers, $\Delta\mu$ should be of the same order as the barrier to internal rotation (*17*). To compare the polymers studied we take a value of approximately 2 K cal/mol for rotation about the C–C bond (*39*) and about 1 K cal/mol for rotation about the Si–O bond (*40*) as a basis for no more than a rough estimate of Z*. Activation energies have been measured at widely differing temperatures, and it is instructive to compare estimates of Z* reduced to a common temperature. Following Adam and Gibbs (*17*) the ratio of the values of Z* at two temperatures, T' and T'', may be expressed in terms of the ratio of configurational entropies as:

$$\frac{Z^*(T')}{Z^*(T'')} = \frac{S_c(T'')}{S_c(T')} = \frac{\ln(T''/T_2)}{\ln(T'/T_2)} \tag{14}$$

where T_2 is the temperature at which configurational entropy vanishes. Using Equation 14 and published values for T_2 (PPO, 112 K; PIB, 132 K (*17*)) we have reduced Z* to the common temperature of 237.5 K, the midpoint of the PDMS measurement range. Estimates of Z* based on the values chosen for $\Delta\mu$, and as reduced are tabulated below:

	Temperature (Range of Measurement)	*Z*(ΔH/Δμ)*	*Z*(237.5 K)*
PDMS	235–240 K	11	11
PIB	430–460 K	11	22
PPO	300–320 K	10	14

These values appear reasonable, but in the absence of any independent assessment of the degree of cooperativity they do not constitute a test of the model. Indeed the measured values of activation energy could be accounted for reasonably by models such as the free-volume model of Bueche (*41*). On the other hand, it is difficult to account for the behavior of the cross-linked gums without invoking a cooperative description.

Relaxation in Networks. Any model which considered the individual chains between cross-links would predict faster relaxation as the free chain lengths are shortened—that is, as the rubber is stiffened. This is contrary to our observations. We can account for our data at least qualitatively in terms of the cooperative model as follows: increased cross-link density extends the size of the cooperative regions, lowering relaxation rates according to Equation 13. Well above the hypersonic relaxation temperature (240 K in PDMS) relaxation rates in the gum are faster than the rates of rearrangement that accommodate the Brillouin phonons. Cross-linking reduces relaxation rates until at densities corresponding to approximately 35 m.u. between cross-links (G is approximately equal to 9×10^5 N m^{-2}) relaxation rates correspond to about 2.5 GHz at room temperature giving rise to a loss maximum at this frequency (Figure 5b). With further cross-linking, relaxation rates fall below Brillouin frequencies and the modulus approaches a glassy plateau (Figure 5a). Measured at

5 GHz there is less reduction in relaxation rates, hence less cross-linking is needed to observe the maximum loss, and the glassy plateau is reached sooner. The relationship between relaxation rates and cross-link density will depend on the manner in which Z* increases with cross-linking. In the simplest case Z* might increase in direct proportion to cross-link density, since the entropy of a chain falls in proportion to cross-link density (*42*), so in this approximation

$$Z^* (\rho) = Z^* (o) + B\rho \tag{15}$$

where Z* (o) is the size of the cooperatively rearranging region in the uncross-linked gum and ρ is the cross-link density which we shall take as the reciprocal of the number of monomer units between cross-links. We can obtain a value for the constant B from our data if we assume that relaxation rates are still described by Equation 13 with the value for Z* given by Equation 15:

$$\overline{W}(T) = \bar{A} \exp \left[-\Delta\mu/kT \left(Z^* (o) + B\rho \right) \right] \tag{16}$$

so that we may write the ratio of the relaxation times τ_1/τ_2 at cross-link densities ρ_1 and ρ_2 as:

$$\ln \tau_1/\tau_2 = \frac{B\Delta\mu}{kT} (\rho_1 - \rho_2) \tag{17}$$

At 293 K we have $\tau = (2.5 \times 10^9)^{-1}$ at $\rho = 1/35$ and $\tau = (5 \times 10^9)^{-1}$ at $\rho - 1/84$. Using $\Delta\mu = 1$ K cal gives a value for B of 24 m.u. At the cross-link densities used this value for B would imply a change of only a few percent in Z* owing to cross-linking, not enough to move the major hypersonic loss up to room temperature from 240 K. (We may estimate the required change in Z* using Equation 14 and estimating T_2 for PDMS as $T_g - 50°$ approximately equal to 100 K.) So while this model accounts for the features observed it is not quantitatively consistent with the view that the α relaxation is moved upward in temperature by cross-linking in a manner analogous to the movement of the quasi-static glass transition with temperature (*35*). What effect, if any, cross-linking has on the major hypersonic loss awaits the outcome of low-temperature measurements on these networks.

Multiple Relaxation and the Assignment of Loss Peaks. An unambiguous assignment of the hypersonic loss is important if Brillouin measurements are to shed light on the nature of the quasi-static glass transition. Is the hypersonic transition simply an observation of the glass transition at high frequencies or is some additional high-frequency phenomenon being observed, bearing no direct relationship to the glass transition?

The transition map indicates a direct connection with the glass transition. However, the connection is confused by the merging of the α and β lines at high frequencies. The one universal theory of the β relaxation predicts the disappearance of the transition at high temperatures so that the major hypersonic loss process would be caused by the α relaxation alone (*43*). (Incidentally the β relaxation is predicted by this theory to be a consequence of the cooperative nature of the glass transition, and is obtained by introducing a cooperative coupling to the two-level system that separates the free and frozen segments). However, Patterson (*30*) has reported the observation of a further minor hypersonic loss peak at 370 K in PPO which fits a distinct curve for the α relaxation obtained from further high-frequency NMR and dielectric data. Although no further high-frequency data is available for the β relaxation, the 320 K loss peak does fit an extrapolation of the β loss line.

If this assignment is correct it is surprising that the α process should be so much weaker than the β process. It is a result which would cast doubt on the admittedly limited understanding of the β process (*43*). Faced with this uncertainty in the assignment of the major hypersonic loss it seems fruitless to comment on the apparent structure within the peak, except to mention that it has been suggested that this may arise as a consequence of critical fluctuations (*44*).

Likewise our assignment of the major loss peak in PDMS may be subject to some uncertainty. In supporting our original assignment (*16*) we made use of microwave loss data (*45*). It may be that the microwave loss is collision induced (*46*) rather than caused by backbone motion, so that the loss peak at 240 K may correspond to the β relaxation, while the rather weaker loss peak at about 320 K may correspond to the α relaxation (*47*).

In PIB degradation of the sample prevents observation of hypersonic losses at higher temperatures.

Conclusions

Brillouin measurements to date indicate the need for a cooperative model of polymer motion. The relationship between polymer dynamics at high and low temperatures is obscured by the difficulty in making unambiguous assignments of hypersonic loss peaks and the limited understanding of the β process. While some understanding of the high-frequency dynamics has been obtained it appears that an important structural inference may be obtained from Brillouin studies—that is that packing in the cooperatively rearranging regions is indeed amorphous. Further insight into the cooperative motion of polymer chains may be gained by extending studies of the behavior of chains constrained by cross-links, and this work is in progress.

The use of Brillouin scattering in polymer physics is comparatively new—with the time the technique could add greatly to our understanding of polymer dynamics. We hope we have demonstrated some of this potential in this chapter.

Acknowledgments

We are indebted to C. G. Delides for help in preparation of the network samples and to A. Adshead for assistance in obtaining experimental data.

Note

Since the original presentation of this chapter in August, 1977 Ian Shepherd died tragically, shortly after taking up a new post in America. I would like to record the loss of his friendship and his guidance felt by all of his former colleagues in Manchester and elsewhere, and to pay tribute to his pioneering work in the field of laser spectroscopy of polymers. (S. M. Lindsay)

Literature Cited

1. Sandorcock, J. R., *Opt. Commun.* (1970) **2,** 73.
2. Patterson, G. D., Latham, J. P., "Brillouin Scattering from Polymers," *Macromol. Rev.,* in press.
3. Romberger, A. B., Eastman, D. P., Hunt, J. L., *J. Chem. Phys.* (1969) **51,** 3723.
4. Friedman, E. A., Ritger, A. J., Andrews, R. D., *J. Appl. Phys.* (1969) **40,** 4243.
5. Jackson, D. A., Pentecost, H. T. A., Powles, J. G., *Mol. Phys.* (1972) **23,** 425.
6. Mitchell, R. S., Guillet, J. E., *J. Polym. Sci., Polym. Chem. Ed.* (1974) **12,** 713.
7. Peticolas, W. L., Stegeman, G. A., Stoicheff, B. P., *Phys. Rev. Lett.* (1967) **18,** 1130.
8. Friedman, E. A., Ritger, A. J., Huang, Y. Y., *Bull. Am. Phys. Soc.* (1970) **15,** 282.
9. Gayles, J. N., Peticolas, W. L., *Light Scattering Spectra Solids, Proc. Int. Conf.* (1969) 715.
10. Brody, E. M., Lubell, C. J., Beatty, L. L., *J. Polym. Sci., Polym. Phys. Ed.* (1975) **13,** 295.
11. Lindsay, S. M., Hartley, A. J., Shepherd, I. W., *Polymer* (1976) **17,** 50.
12. Huang, Y. Y., Wang, C. H., *J. Chem. Phys.* (1975) **62,** 120.
13. Ibid. (1974) **61,** 1868.
14. Huang, Y. Y., et al., *Proc. Int. Conf. Light Scattering Solids, 2nd, 1971,* 488.
15. Patterson, G. D., *J. Polym. Sci. A2* (1977) **15,** 455.
16. Lindsay, S. M., Adshead, A., Shepherd, I. W., *Polymer* (1977) **18,** 862.
17. Adam, G., Gibbs, J. H., *J. Chem. Phys.* (1965) **43,** 139.
18. Leontovitch, M., *Z. Phys.* (1931) **72,** 247.
19. Lindsay, S. M., Shepherd, I. W., unpublished data.

20. Montrose, C. J., Solovyev, V. A., Litovitz, T. A., *J. Acoust. Soc. Am.* (1968) **43,** 117.
21. Lindsay, S. M., Ph.D. Thesis, University of Manchester (1976).
22. Lindsay, S. M., Shepherd, I. W., *J. Phys. E.* (1977) **10,** 150.
23. Lindsay, S. M., Burgess, S., Shepherd, I. W., *Appl. Opt.* (1977) **16,** 1404.
24. Delides, C. G., Shepherd, I. W., *Radiat. Phys. Chem.* (1977) **10,** 379.
25. Waters, A., *J. Appl. Phys.* (1965) **16,** 557.
26. Danielmeyer, H. G., *J. Accoust. Soc. Am.* (1970) **47,** 151.
27. Litovitz, T. A., Davies, C. M., "Structural Relaxation," *Phys. Acoust.* (1965) **2A,** 281.
28. Pinnow, D. A., Candau, S. J., LaMacchia, J. T., Litovitz, T. A., *J. Acoust. Soc. Am.* (1968) **43,** 131.
29. Barlow, A. J., Erginsav, A., *Polymer* (1975) **16,** 110.
30. Patterson, G. D., Douglass, D. C., Latham, J. P., *Macromolecules* (1978) **11**(1), 263.
31. Herzfeld, K. F., Litovitz, T. A., "Absorption and Dispersion of Ultrasonic Waves," Academic, New York, 1959.
32. McCrum, N. G., Read, B. E., Williams, G., "Anelastic and Dielectric Effects in Polymeric Solids," p. 293, Wiley, London, 1967.
33. Ibid, p. 573.
34. Williams, J., Shohamy, E., Reich, S., Eisenberg, A., *Phys. Rev. Lett.* (1975) **35,** 951.
35. Lindsay, S. M., Delides, C. G., Adshead, A., Shepherd, I. W., "Fabry Perot Spectroscopy of Siloxane Networks," presented at the meeting of the Networks Club, Imperial College, London, 1977.
36. Charlesby, A., *Proc. R. Soc., London, Ser. A* (1952) **215,** 187.
37. Berry, G. C., Fox, T. G., *Adv. Polym. Sci. (Fortschr. Hochpolym. Forsch.)* (1968) **5,** 261.
38. Vacher, R., Pelous, J., *Phys. Lett.* (1976) **58A,** 139.
39. Flory, P. J., "Statistical Mechanics of Chain Molecules," Chap. III, Interscience, New York, 1969.
40. Maxfield, J., Shepherd, I. W., *Chem. Phys.* (1973) **2,** 433.
41. Bueche, F., *J. Chem. Phys.* (1959) **30**(3), 748.
42. Treloar, L. R. G., "The Physics of Rubber Elasticity," Clarendon, Oxford, 1949.
43. Brereton, M. G., Davies, G. P., *Polymer* (1977) **18,** 764.
44. Professor S. F. Edwards, private communication.
45. Dasgupta, A., Smyth, C. A., *J. Chem. Phys.* (1967) **47,** 2911.
46. Stumper, U., "Molecular Motions in Liquids," p. 655, Lascombe, J., Ed., Holland, Reidel, 1974.
47. Patterson, G. D., private communication.

RECEIVED November 10, 1977.

12

Ion Recombination Luminescence of Heterogeneous Polymer Systems

G. G. A. BÖHM and K. R. LUCAS

Firestone Tire & Rubber Co., Central Research Laboratories, Akron, OH 44317

Radioluminescence spectroscopy has been used to examine molecular motion, solubility, and morphology of heterogeneous polymer blends and block copolymers. The molecular processes involved in the origin of luminescence are described for simple blends and for complicated systems with interphases. A relatively miscible blend of polybutadiene (PBD) and poly(butadiene-co-styrene) and an immiscible blend of PBD and EPDM are examined. Selective tagging of one of the polymers with chromophores in combination with a spectral analysis of the light given off at the luminescence maxima gives quantitative information on the solubility of the blend components in each other. Finally, it is possible to substantiate the existence and to measure the volume contribution of an interphase in styrene–butadiene–styrene block copolymers.

Luminescence emission from irradiated materials has been known for many years. The phenomenon has been studied primarily on inorganic systems but also with polymers. A review of this work has been published by Partridge (*1*). A clearer understanding of the processes involved has led in more recent years to the use of this technique as a tool to probe the structure of heterogeneous polymer systems (*2, 3*). This chapter describes the technique and shows its utility for the analysis of polymer blends and block copolymers.

Processes Governing Radioluminescence Emission

The method involves measuring and analyzing light given off by a polymer following exposure to ionizing radiation. Most commonly the polymer is irradiated at a temperature considerably below its glass

0-8412-0406-3/79/33-174-227$05.00/1

transition temperature (T_g), and the emitted light is measured during warm-up to $T > T_g$. The equipment used consists of a cryogenically cooled sample holder protruding into a vacuum chamber to which the irradiation source and the optical equipment are attached. A monochromator and photomultiplier provide a spectral analysis of the light given off by the sample. A more detailed description of the equipment was published elsewhere (*4*). The molecular processes involved in radioluminescence spectroscopy are discussed below.

Homopolymers. When a high-energy electron is traversing a polymer sample, secondary electrons are produced along its track. These electrons possess considerable kinetic energy which they dissipate by ionizing and exciting molecules immediately surrounding the parent ion. This interaction with the matrix limits the range of the secondary electrons and defines the dimensions of the cage in which the reaction products are confined.

If the irradiation of the polymer is carried out above the T_g, the secondary electron will, after having spent its kinetic energy, be drawn back to and recombine with the parent ion or any other positive ion in the vicinity. Irradiation of the polymer in the glassy state, in contrast, will lead to a trapping of some of the secondary electrons by neutral molecules with positive electron affinity, free radicals, dielectric cavities,

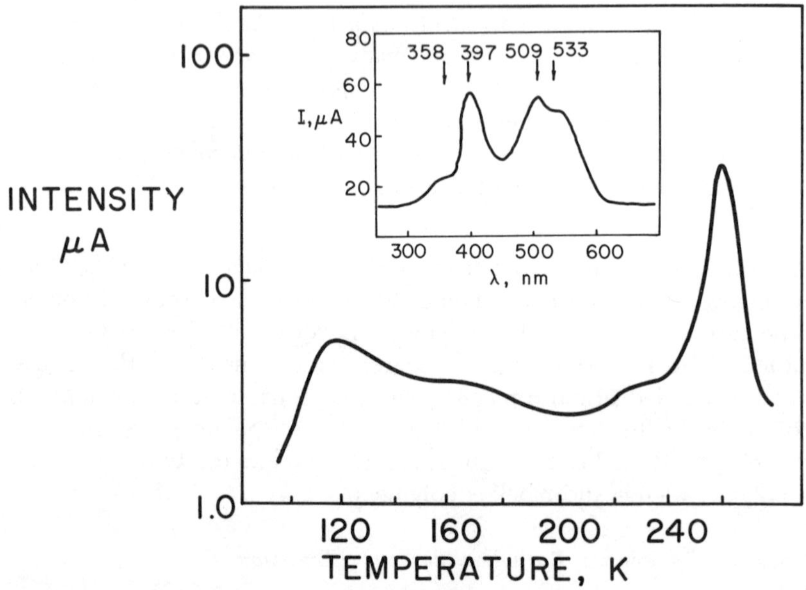

Figure 1. Luminescence of PBD irradiated with electrons at 90 K to a dose of 3.1 · 10^{19} eV g^{-1}. Insert shows spectral distribution of light given off at 258 K.

etc. They are released from these traps very slowly on isothermal storage at $T << T_g$ or more rapidly on warm-up of the sample to $T > T_g$. In the latter case, an erosion of the traps is induced by the onset of local motion in the polymer involving molecules located in the immediate vicinity of the electron trap. In cases where an electron is attached to a species of high electron affinity such as a radical, it is likely that the lattice energies kT at $T < T_g$ are not large enough to free the electron from its bondage. Its contribution to the luminescence of the polymer will thus be delayed until the sample undergoes a transition from the glass to the liquid state. At that temperature long range motion sets in, and charge neutralization occurs by physical approach and reaction of the radical ion R^- and one of the M^+ ions. We believe this process rather than electron detrapping is responsible for the strong luminescence maxima observed in many polymers at their respective T_g.

Regardless of whether or not the secondary electron is temporarily trapped, it will in time recombine with a positive ion, and charge neutrality will again be established. This event, as well as the reactions induced by the electron during its travel through the matrix will lead to a proliferation of stable and unstable products. Some of the electronically excited molecules produced in this manner will return to their ground state by luminescence emission and by inter- or intramolecular energy transfer. The photons given off by the sample can thus come from the excited molecule or from chemicals attached to or contained in the polymer.

Summarizing then, we can state that the luminescence emitted by an irradiated polymer during warm-up from $T << T_g$ to T_g (glow curve) will exhibit maxima at temperatures corresponding to molecular relaxation processes commencing in the polymer below and at T_g. The former might be caused by the onset of motion involving specific parts of the molecule such as side groups or a rearrangement of whole segments of the polymer molecule in some local regions in a manner proposed for molecules lacking internal modes of motion (*5*). An example is given in Figure 1 showing the glow curve for polybutadiene along with the spectral distribution of light measured at T_g.

The intensity of light measured at a particular temperature is proportional to the rate of electron recombination:

$$I \propto \phi \frac{dn}{dt} \tag{1}$$

A fraction ϕ of the electrons will cause an emission of light. Most electrons contributing to the luminescence of the irradiated polymer are mobilized through an erosion of cavity traps. A fraction, however, is generated by light-induced detrapping (photo-bleaching), a process which has been described by Böhm (*4*). Photons produced by the former process cause the liberation from traps of more electrons which, in turn, can cause luminescence. The secondary photons will bleach more electrons and so forth.

The total luminescence intensity is then:

$$I(T) = \left(\frac{dn}{dt}\right)_{\mathrm{e}} (\phi + k\phi^2 + k^2\phi^3 + \ldots) \tag{2}$$

where $(dn/dt)_{\mathrm{e}}$ is the rate of electron mobilization by erosion of traps and k the probability that a photon will remove an electron from its trap. The bleaching at a temperature T will free electrons from traps deeper or of greater electron affinity than those which are eroded at T. Hence, the more bleaching occurs at or before T, the smaller the luminescence intensities at temperatures above T. The overall contribution of bleached electrons to the luminescence given off by the polymer should be relatively small. This one can reason from the fact that ϕ and k are both smaller than one.

Polymer Blends. Let us now apply this technique to the study of polymer blends. We will assume that the two materials have different sets of (mechanical) transition temperatures and that energy spectra of the luminescent light given off by the two polymers are distinguishably different. If the two polymers are completely miscible on a molecular scale (a rare event indeed), we can expect that many of the luminescence maxima will be seen at temperatures different from those of the blend components. This will certainly apply to the T_{g} of the miscible blend which can be approximated by:

$$T_{\mathrm{g}}(\text{blend}) = X_{\mathrm{n}}{}^{\mathrm{A}}T_{\mathrm{g}}{}^{\mathrm{A}} + X_{\mathrm{n}}{}^{\mathrm{B}}T_{\mathrm{g}}{}^{\mathrm{B}} \tag{3}$$

where $X_n{}^{\mathrm{A}}$ and $X_{\mathrm{n}}{}^{\mathrm{B}}$ are the volume fractions and $T_{\mathrm{g}}{}^{\mathrm{A}}$ and $T_{\mathrm{g}}{}^{\mathrm{B}}$ are the glass transition temperatures of the two polymers in the blend. The energy spectrum of the light given off at the new T_{g} (blend) will comprise the spectra of both components with an intensity ratio governed by the luminescence efficiency of the polymers and by energy transfer processes occurring in the blend.

The second more common case is represented by a blend in which the two polymers are not miscible. The resulting heterogeneous system will yield a luminescence spectrum different from that of a miscible blend. This can be rationalized using Figure 2 which is a schematic of a region near the interface between domains of the phase-separated blend com-

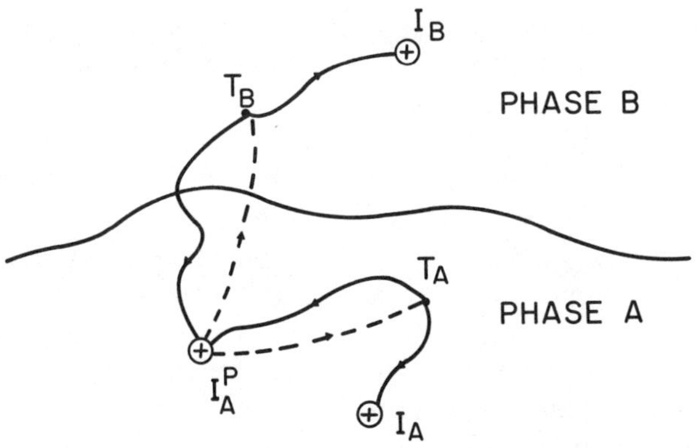

Figure 2. Ion recombination processes. Schematic of electron trapping in heterogeneous polymer blend near domain interface.

ponents, polymers A and B. It is assumed that both polymers were cooled to the glassy state. Then on radiation exposure a secondary electron generated in and travelling through the frozen matrix A will most likely be trapped at T_A in phase A. If the secondary electron is generated near the AB phase boundary it can, however, cross the interface and be trapped at T_B in phase B. The probability for this event will depend on the size and shape of the domains in the heterogeneous blend and on the average travel range of secondary electrons. It has been estimated that most of the secondary electrons produced have energies below the ionization potential of common molecules, and hence their range in polymers of unit density should be only a few angstroms. However, a small percentage of the electrons produced can possess energies of up to several hundred eV. These so-called δ electrons can have spur diameters of 50 Å and more (*6*). One can make a similar argument for the trapping of secondary electrons generated in phase B.

We shall now address ourselves to the fate of the secondary electrons trapped in phases A and B following their release from these traps on warm-up of the sample to $T > T_g{}^A$ and $T_g{}^B$. The electrons trapped in phase A should more or less all combine with ions present in the same phase. Very likely they will combine with the parent ion $I_A{}^P$ from which the electron originated or with an ion I_A belonging to a cluster of ions formed along with the parent ion by a δ electron of higher energy. Those electrons which transferred and were trapped in phase B could in principle recombine with ion I_B of the B phase. Again, however, it is probable that most of these electrons, attracted by the Coulomb field of their parent ions or of ions of the parent spur, will return to phase A and recombine there. The reason for this is that once an electron is released from one of the traps, its energy is only kT. Without much own initiative, its travel and range is thus governed by the Coulomb field of ions in the vicinity. The separation between positive and negative ions required to reduce the energy of Coulomb attraction to the thermal energy kT and $T = 77$ K is about 1000 Å. Remembering that the average travel distance of a secondary electron following irradiation is less than 50 Å compared with an average distance between two ions of about 200 Å (based on a random distribution of trapped ions generated with a G value of G (trapped ion) ~ 0.1 (7) and a dose of 2 Mrad), we can conclude that the parent ion, on the average, will be the closest neighbor to the trapped electron. Hence, it is likely to recombine with it. This behavior is different from that observed with frozen organic compounds such as 3-methylpentane. There experiments with electron scavengers such as biphenyl have shown that a considerable portion of the trapped electrons are not constrained to combine with a predestined positive ion. The difference we attribute to the much greater electron trapping probability for low-molecular-weight hydrocarbon glasses ($G(e) \sim 1$) compared with the values for polymers.

Our hypothesis thus assumes that at T greater than T_g^A and T_g^B, essentially all the electrons created in phase A will have returned to A, and most electrons created in B will have returned to B. Yet the luminescence spectrum of the blend may not be that of A plus B. This is because the electrons crossing the phase boundary from B to A will do so at a transition temperature characteristic of polymer B, but the light given off on their recombination with ions of the A phase will have the energy spectrum of A rather than of B. Thus, by measuring the energy spectrum of the luminescent light given off at the transition temperatures of A and B, the number of electrons transferring from A to B and from B to A during warm-up of the irradiated blend can be estimated. From it, conclusions can be drawn about the size of the domains.

Heterogeneous Polymer Systems with Interphase Regions. Radioluminescence measurements can also be used to analyze interphases, that is, regions in a polymer which separate the domains of a heterogeneous system. The volume fraction of material in the interphase region in principle can be estimated by thermodynamic calculations and should depend on such factors as the relative solubility of the polymeric blend components and whether or not a chemical link exists between them such as is the case in block copolymers, etc. The attainment of a thermodynamic equilibrium, however, is often prevented by kinetics. That is, if a particular interphase configuration is arrived at by processing of the polymer, e.g., high shear mixing, etc., the system may find it difficult or impossible to assume a morphology prescribed by thermodynamics because of the slow viscoelastic response of the macromolecules. In these cases, a determination of the interphase region will have to rely on suitable experiments rather than on thermodynamic calculations.

Information on interphases in polymers is rather scarce. Most studies dealt with styrene–isoprene–styrene (SIS) block copolymers for which an interphase was predicted by theory (8) and is measured by small-angle x-ray scattering (9).

The glow curve of such a heterogeneous system we expect to be the sum of the contributions from regions pure in A and B as well as from the interphase:

$$I(T) = X_A I_A(T) + X_B I_B(T) + \sum_1^n X_m I_m(T) \quad (4)$$

For the purpose of estimating the glow curve of the interphase, one can think of it as an ensemble on n regions of composition X_m^A. Regions of equal composition may exist in a heterogeneous polymer system at different locations. However, regardless of where they are relative to each other, they should exhibit the same type of glow curve. Hence, one is justified in writing the joint contribution of all these regions to the overall luminescence intensity as $X_m I_m(T)$ where X_m is the volume fraction of material in all these regions and $I_m(T)$ is the normalized luminescence intensity of a blend with composition X_m^A. The total contribution of the interphase is then obviously the sum of the contributions from the entire ensemble of regions as expressed in Equation 4. To carry out this summation, we need to know the glow curves of miscible blends in the composition range of interest. These can often be determined in experiments. They can also be estimated from the corresponding curves for the component polymers.

The glow curve of a miscible blend will have characteristic features similar to those of polymer A and B; the intensity maxima, however, may be seen at different temperatures. The major transition will now appear at the T_g of the blend defined by Equation 3. The intensity of light given off on passing through this maximum should then be the sum over the weighted intensity contributions of the pure blend components measured at their respective T_g. The low intensity maxima present in the glow curves of the pure polymers A and B are expected to be retained in the blend without an appreciable shift of the transition temperatures. The above then suggests that the total contribution of the interphase can be calculated by:

$$I^{IP}(T) = \sum_1^n X_m[X_m^A I^A(T') + X_m^B I^B(T')] \quad (5)$$

following a transposition of the glow curves $I(T)$ for the polymers A and B to $I(T')$ by a transformation of the temperature coordinate according to:

$$T' = T + (T - T_\mathrm{I})k^{\mathrm{A,B}}$$
$$k^{\mathrm{A,B}} = (T_\mathrm{g}(m) - T_\mathrm{g}^{\mathrm{A,B}})/(T_\mathrm{g}^{\mathrm{A,B}} - T_\mathrm{I}) \qquad (6)$$

where T_I is the irradiation temperature. Since the total number of photons emitted should not be altered by the transposition of the glow curves, we need to introduce an intensity correction function. For this, we chose:

$$I(T') = f(T - T_\mathrm{I})I(T) \qquad (7)$$

where f is determined from $\int_{T_\mathrm{I}}^{T > T_g^{\mathrm{A,B}}} I(T)\,\mathrm{d}T = \int_{T_\mathrm{I}}^{T > T_g^{\mathrm{A,B}}} I(T')\,\mathrm{d}T'$ (8)

Finally, we need to keep track of our material balance in the pure phases and in the interphase which is governed by the following equations:

$$X_\mathrm{A}^\circ + X_\mathrm{B}^\circ = 1 \qquad (9)$$

$$X_\mathrm{A} = X_\mathrm{A}^\circ - X_\mathrm{A}^{\mathrm{IP}}, \quad X_\mathrm{B} = X_\mathrm{B}^\circ - X_\mathrm{B}^{\mathrm{IP}} \qquad (10)$$

$$X^{\mathrm{IP}} = \sum_1^n X_\mathrm{m} = X_\mathrm{A}^{\mathrm{IP}} + X_\mathrm{B}^{\mathrm{IP}} \qquad (11)$$

where X_A°, X_A, and $X_\mathrm{A}^{\mathrm{IP}}$ are the volume fractions of A in the entire sample, in the regions pure in A, and in the interphase, respectively. X^{IP} and X_m are the volume fractions of the interphase and of the m'th part of the ensemble of interphase regions having a composition X_m^A, X_m^B.

Using Equations 4–11, it should now be possible to determine critical information on the interphase region such as X^{IP} and X_m^A from known values of X_A° and experimentally determined glow curves of the heterogeneous system and of the component polymers contained in it. A knowledge of the radiation-induced changes occurring in a specific pair of component polymers may suggest a modification of the above-discussed analysis method, particularly in Equation 5, to improve on the accuracy of the technique.

Examples

Blends of PBD and Poly(butadiene-co-styrene). An example of the luminescence spectrum of a blend is shown in Figure 3. There data is presented for a composite comprised of 50 wt % PBD and 50 wt % poly(butadiene-co-styrene) before and after annealing alongside with the spectra of the blend components. The blends were prepared by mixing in a Brabender plasticorder followed by milling. The blends tested shortly after preparation exhibit transitions corresponding to the T_g peaks of the two component polymers. After annealing the sample at 423 K in vacuo for 26 hr, a different spectrum is obtained. Instead of the two dominant peaks at 178 and 202 K, we notice a strong peak at 191 K and on its shoulder a less pronounced transition at about 197 K. This finding indicates a considerably increased miscibility of the two polymers at the annealing temperature. This may have been caused by a reduction in the free energy of mixing on increase in temperature as

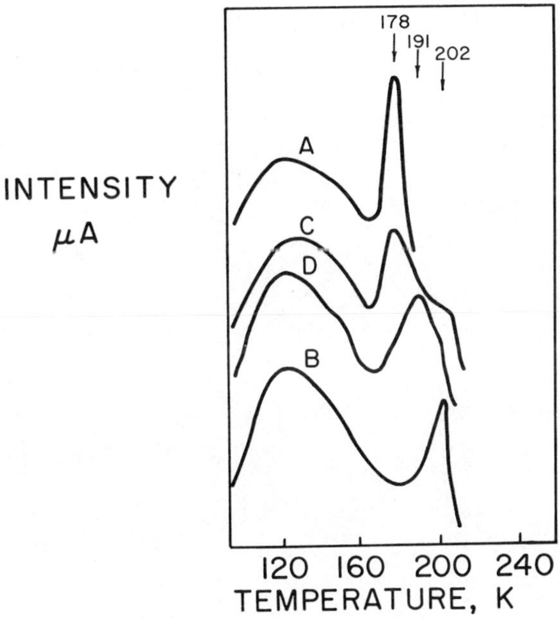

Figure 3. Luminescence of irradiated polymers: (A), PBD (cis/trans/vinyl = 55/35/11); (B), (polybutadiene-co-styrene) with 20% styrene; (C, D), blend comprising 50 wt % of each of the two elastomers prior to and after annealing in vacuo at 423 K for 26 hr. Electron irradiation at 90 K to a dose of $3.1 \cdot 10^{19}$ eV g^{-1}.

might be expected from operation in an upper critical solution temperature regime, or it could be that a more miscible state at 295 K could not be reached because of kinetics.

Blends of Ethylene/Propylene/Ethylidene Norbornene Terpolymer (EPDM) and PBD. A more detailed analysis of a blend system can be made if a spectral analysis is made of the light given off by the sample at the various intensity maxima. This can be illustrated in Figures 4–7 showing results obtained with a 50/50 blend of EPDM ($M_n = 105 \cdot 10^3$, $M_w/M_n = 3$) and PBD ($M_n = 172 \cdot 10^3$, $M_w/M_n = 1.8$, 81% vinyl unsat.) To obtain a strong and clearly discernible spectral response from both blend components, the EPDM was labeled with mercaptoanthracene. The latter was grafted onto the EPDM by reaction with the double bonds of the ethylidene norbornene contained in the ethylene propylene rubber. The energy spectrum of the light emitted from this tagged polymer (Figure 4) shows a single, relatively broad peak with a maximum at 430 nm which is well discernible from the four-peak spectrum of PBD (Figure 5) with maxima at 330, 390, 500, and 525 nm.

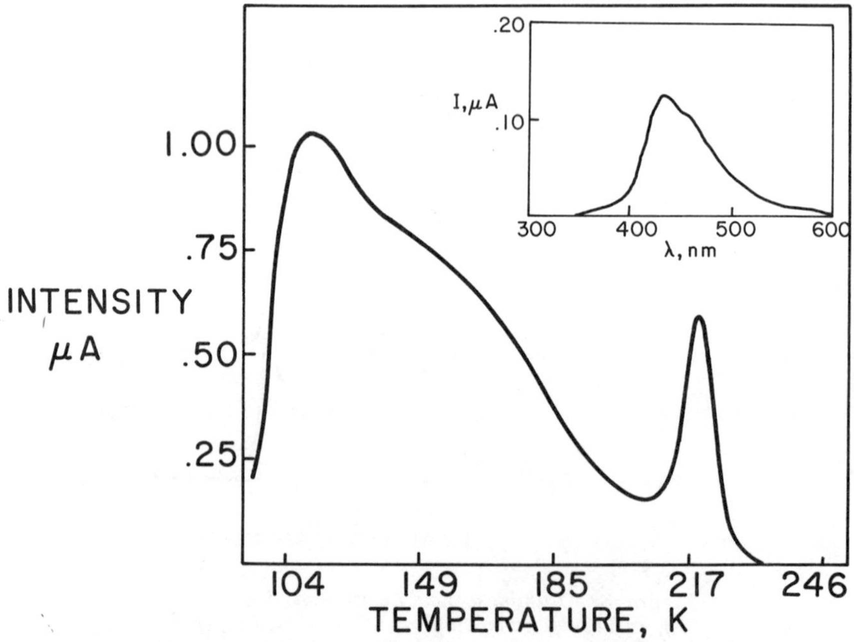

Figure 4. Luminescence of mercaptoanthracene tagged ethylene/propylene/ethylidene norbornene terpolymer irradiated with x-rays at 90 K to a dose of $5 \cdot 10^{19}$ eV g^{-1}. Insert shows spectral distribution of light given off at 220 K.

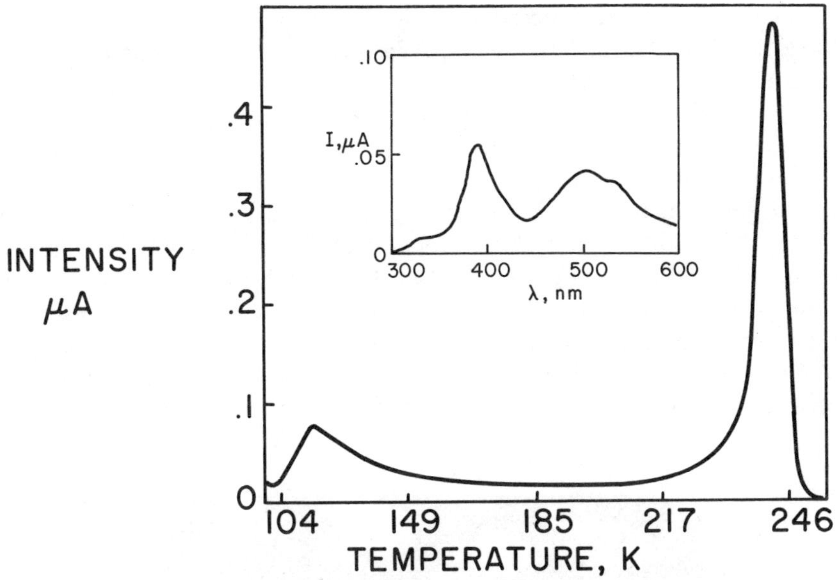

Figure 5. Luminescence of PBD irradiated with x-rays at 90 K to a dose of $5 \cdot 10^{19}$ eV g^{-1}. Insert shows spectral distribution of light given off at 240 K.

A blend comprising equal parts of the tagged EPDM and PBD should be heterogeneous in nature. This one can anticipate from the considerable difference of the solubility parameters and the high molecular weight of the two polymers. A photomicrograph taken with a transmission electron microscope confirms this. Figure 6 shows discrete domains of EPDM approximately 1.5 μm in size embedded in a continuous matrix of PBD. The latter was stained by the OsO_4 and hence appears black. The interface between the domains appears sharp suggesting the absence of an appreciable interphase region. The luminescence given off by the irradiated blend sample is shown in Figure 7. The peaks of the glow curve at temperatures of 220 and 240 K correspond to the strong intensity maxima recorded for the two component polymers at their respective T_g (Figures 4 and 5). A very slight shift in the temperatures at which the T_g (EPDM) maximum occurs in the blend relative to the pure state suggests a very small miscibility of PBD in EPDM. Using Equation 3, one can correlate the observed shifts of ΔT(EPDM) = 1.3 K with a miscibility of 6.5% PBD in EPDM. No shift was observed of the peak at T_g(PBD). The width of the two main peaks observed for the blend are about equal to those of the pure materials, hence one can conclude that no appreciable interphase exists between the domains of the component polymers. These conclusions are verified by the energy spectrum of the light given off by the irradiated blend at $T = 240$ K (Figure 7). Within the limit of accuracy of this measurement the spectrum is that of pure PBD since there is no evidence for the presence of the strong single peak emission spectrum of EPDM at 430 nm. In contrast, a contribution of both EPDM and PBD is noted in the spectrum measured at $T = 220$ K. This must in part be caused by the small amount of PBD dissolved in EPDM, but another reason for this is that a detrapping of electrons located in the glassy PBD phase occurs induced by the light emitted in the EPDM domains while passing through the glass transition region at this temperature. The resulting light emission from the glassy PBD then contributes to the overall luminescence measured at $T = 220$ K. The detrapping in polymer glasses of electrons by visible light has been discussed briefly earlier. Finally,

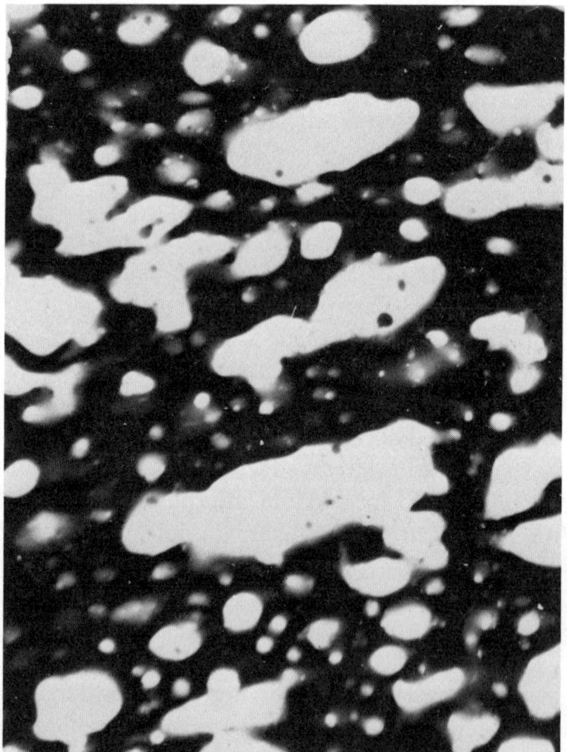

Figure 6. Transmission electron micrograph of OsO_4 stained blend of 50 wt % PBD and 50 wt % ethylene/propylene/ethylidene norbornene terpolymer. Magnification, 3300×.

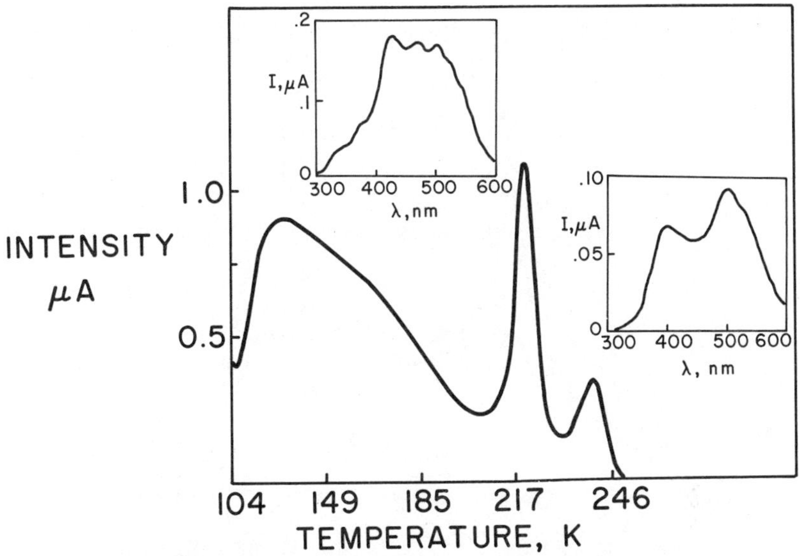

Figure 7. Luminescence of blend containing 50 wt % PBD and 50 wt % ethylene/propylene/ethylidene norbornene terpolymer irradiated with x-rays at 90 K to a dose of $5 \cdot 10^{19}$ eV g^{-1}. Inserts show spectral distribution of light emitted at 221.5 and 240 K, respectively.

the intensity of the peak at T_g(PBD) is smaller than would be expected from the material balance referred to above. The decrease we attribute to triplet state quenching by traces of O_2 dissolved in the polymer. Evacuation over days will restore the peak height to the correct level.

Butadiene–Styrene Block Copolymers. Our last example is a polymer for which the existence of an interphase between the styrene domains and the PBD matrix has been postulated (8) and experimentally verified (9). Figure 8 shows the luminescence spectrum of a SBS block copolymer (28% styrene, mol wt $= 10^4/5 \cdot 10^4/10^4$) along with the spectra of polystyrene (PS) and PBD. The latter two polymers were similar in microstructure and molecular weight to the PBD and PS branches of the block copolymer. In comparing the T_g-related peak of PBD with the corresponding maximum of the SBS spectrum, one notices a strongly asymmetric shape of the latter peak resulting from a broad tail on the high temperature side of the peak. This suggests the existence of electron traps in the block copolymer that are more stable than those of PBD. Hence, electrons will be freed from their traps and will cause luminescence at $T > T_g$ (PBD) when all the electron recombination processes have long been completed in the PBD phase. The question then remains where in the block copolymer are there traps of such higher stability. As one can see from the luminescence spectrum of pure PS, there

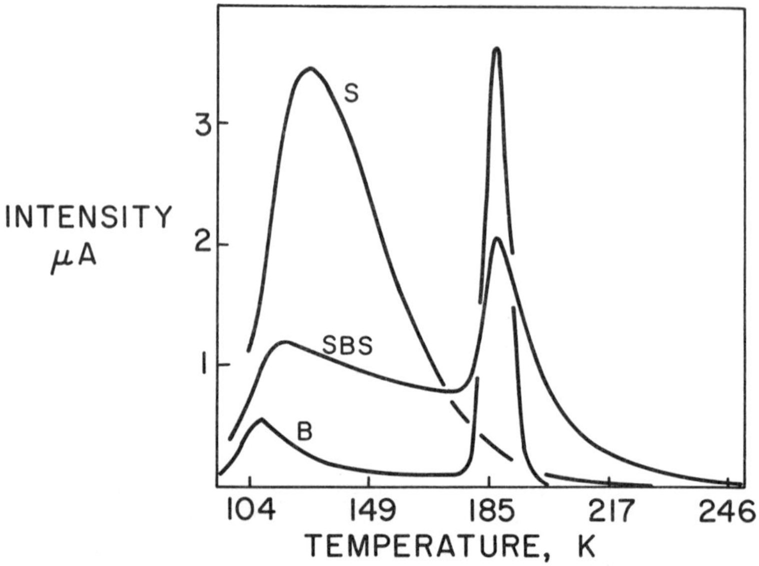

Figure 8. Luminescence of SBS block copolymer as well as of PS and PBD homopolymers similar in microstructure and molecular weight to the component polymers of SBS. Samples were irradiated with x-rays at 90 K to $5 \cdot 10^{19}$ eV g^{-1}.

is no transition occurring near that of the PBD phase in SBS. In fact, appreciable local mobility exists in polystyrene at temperatures considerably below its T_g at 353 K causing a complete erosion of all electron traps in the polymer well before long range motion sets in. Therefore, the light given off by PS at about 190 K is of relatively low intensity since much luminescence had already occurred at lower temperatures. The presence of deeper, more stable traps must thus be sought in a region of the block copolymer which, comprising both PS and PBD, has a T_g above that of PBD. The only place where that could be in this material is the interphase between the PBD and the PS domains. To define the contribution of this region to the overall luminescence given off by the polymer, the PBD- and PS-related light emission was subtracted from the total glow curve after a proper normalization of the glow curves taken for the pure materials. The tail section of this curve shown in Figure 9 (Curve A) extends to a temperature of over 250 K. Using Equation 3 one can calculate that this temperature, defined as the T_g of a particular region of the interphase, corresponds to a composition of approximately 53% PS. It can be assumed that interphase regions with higher styrene content exist in the block copolymer. However, their detection is obscured by the low luminescence emission at these temperatures. This prevents us from making a complete analysis of the interphase region as was outlined earlier. It is possible however to estimate the volume fraction of the interphase X^{IP} by assuming a linear concentration gradient of PBD and PS in this region and by neglecting the luminescence contribution of PS to the light emission of the interphase. The latter seems justified in light of what has been published on the fate of transient species in irradiated PS. Alfimov et al. (*11*) reported that secondary electrons produced at 77 K are trapped mainly by weak interaction with the matrix but also by addition to disubstituted benzyl and cyclohexadienyl radicals formed in PS with a $G(R)$ value of 0.08 (*10*). The former mode of trapping results in the appearance of a narrow ESR singlet signal. Warming the sample to the γ transition temperature or photo-bleaching with visible light at 77 K causes this singlet to disappear under luminescence emission. Hence only very few strongly bound electrons remain trapped in PS at $T \leq T_g$ (PBD). The electrons trapped in the interphase at these temperatures are thus very likely bound to the more abundant radicals formed on the PBD molecules. The luminescence emitted from this region is then likely caused by charge neutralization reactions involving radical ions R^- (PBD) and cations M^+ (PBD).

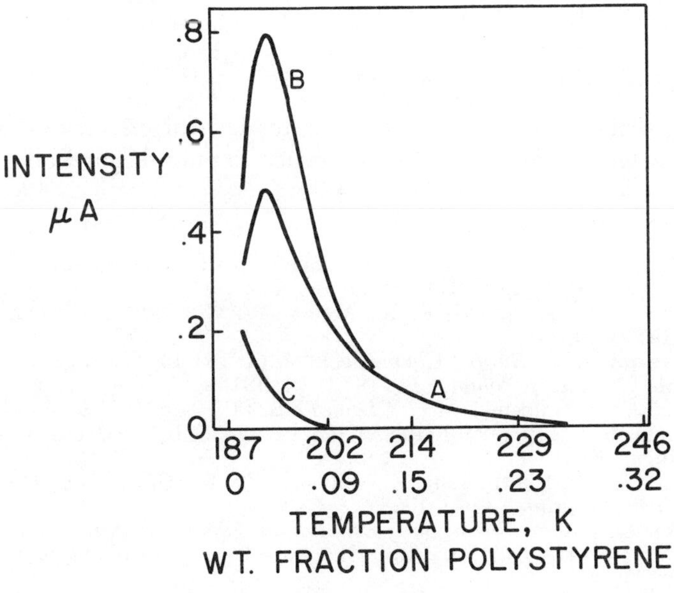

Figure 9. Luminescence contribution of the interphase region present in SBS block copolymer. (A), sample cast from tetrahydrofuran; (B), sample compression molded at 100°C for 10 min; (C), physical blend of 28 wt % PS and 72 wt % PBD.

X^{IP} calculated with the above two assumptions leads to a value of 0.6 for the solvent cast sample. Our data suggests that the quantum yield ϕ is greater for PS than it is for PBD. Hence, if energy transfer to the PS takes place following a charge neutralization process, X^{IP} will be overestimated. Similarly, if the presence of PS reduces $G(R)_{PBD}$, X^{IP} will be underestimated. Hence, to firm up the estimate of X^{IP} it will be necessary to examine these factors more closely.

X^{IP} values for the solvent cast polymer can also be calculated using the theory of Meier (8). We assumed $(\delta_S - \delta_{BD})^2 = 0.6$ and $M = M_n(PS) + 0.5\,M_n(PBD) = 35000$ and obtained $X^{IP} \sim 0.5$. Considering the above referred to uncertainties in the accuracy of the radioluminescence data and the only limited applicability of Meier's theory (the blocks do not have equal length and the concentration gradient is sigmoidal rather than linear as assumed here), one must be surprised about the closeness of the X^{IP} values obtained from theory and experimental data.

Figure 9 also indicates that X^{IP} is larger for a sample molded at 100°C for 10 min compared with that cast from tetrahydrofuran. A comparison of the areas under both curves suggests a ratio of X^{IP}(molded)/X^{IP}(cast) $\sim$ 1.1. An increased X^{IP} for the molded sample would not be surprising since one would not expect thermodynamic equilibrium to be reached under molding conditions.

Acknowledgments

We wish to thank L. E. Vescelius for helpful discussions and The Firestone Tire & Rubber Co. for permission to publish this work.

Literature Cited

1. Partridge, R. H., "The Radiation Chemistry of Macromolecules," Vol. I, Chapter 10, M. Dole, Ed., Academic, New York, 1973.
2. Böhm, G. G. A., Lucas, K. R., Mayes, W. G., *Rubber Chem. Technol.* (1977) **50**, 714.
3. Zlatkevich, L. Y., *Rubber Chem. Technol.* (1976) **49**, 179.
4. Böhm, G. G. A., *J. Polym. Sci.* (1976) **14**, 437.
5. Johari, G. P., Goldstein, M., *J. Chem. Phys.* (1970) **53**(6), 2372.
6. Lea, D. E., "Action of Radiation on Living Cells," 2nd ed., Cambridge University, 1956.
7. Keyser, R. M., Lin, J., Tsuji, K., Williams, F., *Am. Chem. Soc., Div. Polym. Chem. Preprints* (April, 1968).
8. Meier, D. J., *Am. Chem. Soc., Div. Polym. Chem. Preprint* (April, 1974).
9. Hashinioto, T., Todo, A., Itoi, H., Kawai, H., *Macromolecules* (1977) **10**(2), 377.
10. Florin, R. E., Wall, L. A., Brown, D. W., *Trans. Faraday Soc.* (1960) **56**, 1304.
11. Alfimov, M. V., Nikolskii, V. G., Buben, N. Ya., *Kinet. Katal.* (1964) **5**, 268.

RECEIVED November 10, 1977.

13

Structural Analysis by Diffusion Measurements: SBS Block Copolymers and Polyethylene

MICHAEL V. SEFTON and KWAN T. CHIANG

Department of Chemical Engineering and Applied Chemistry, University of Toronto, Toronto, Ontario, M5S 1A4, Canada

The diffusion coefficients of cyclohexane in an SBS block copolymer and of iso-octane in Sclair polyethylene demonstrated unique morphological structures within the polymers. The interphase between polystyrene domains and polybutadiene matrix in SBS TR-41-2443 was identified by the time-dependent nature of the diffusion coefficient. Curves of M_t/M_∞ *plotted against* $t^{1/2}/1$ *for different film thicknesses did not coincide as required by the criteria of Fickian sorption. Furthermore, the presence of internal stress in films of SBS cast on Teflon or glass surfaces has been demonstrated by comparing the apparent diffusion in these films with that measured in films cast on mercury. The diffusion coefficient of iso-octane in Sclair polyethylene unexpectedly increased with increasing crystallinity. This was caused by the presence of fewer, larger crystallites in the more crystalline material, which has a lower chain immobilization factor than the less crystalline material. These experiments, along with examples in the literature, demonstrate the benefits of using transport phenomena to investigate polymer structure.*

Detailed examination of polymer structure can provide polymer engineers with useful information regarding the performance and processing characteristics of the polymer. Conventional techniques (e.g., x-ray diffraction, birefringence) provide excellent analyses of gross morphology but are unable to determine, for example, the presence of internal stress or regions of partly ordered chains. In recent years, attention has

0-8412-0406-3/79/33-174-243$05.00/1

been drawn to the potential use of transport phenomena in polymers to investigate polymer properties since these properties are sensitive to the molecular state and structure of the polymer. This technique amplifies the small morphological changes by displaying large changes in transport properties. Furthermore, because of the nature of the diffusion process, the surface structure may be isolated from that of the bulk. Hence, diffusion measurements provide a simple way to evaluate the structure of the polymer surface, and thus its influence on polymer performance may be more clearly defined.

Diffusion in Polymers

The relationship between the diffusion coefficient and polymer structure has been reviewed in detail in the classic text edited by Crank and Park (*1*). The diffusion coefficient in a polymer is conveniently measured in a sorption experiment. The polymer sample is suspended on a balance and is initially equilibrated with vapor of a given diffusant substance at a certain pressure P_i. It is suddenly exposed to a different pressure P_f of the same vapor, and the rate of absorption of the diffusant is determined. When equilbrium has been attained, the pressure of the diffuant is reduced to P_i, and the rate of desorption is measured. Normally P_i is zero, and the film is initially free of diffusant molecules.

If the polymer is in the form of a thin film of thickness 1 and if the diffusant concentration within the surface of the film is constant, then the diffusion process is governed by:

$$\frac{M_t}{M_\infty} = 4\left(\frac{Dt}{1^2}\right)^{1/2}\left(\frac{1}{\pi^{1/2}} + 2\sum_{n=0}^{\infty}(-1)^n \operatorname{ierfc}\frac{n1}{2(Dt)^{1/2}}\right) \qquad (1)$$

where M_t is the mass of diffusant taken up (or lost) by the sheet in the time t, and M_∞ is the equilibrium amount absorbed (or desorbed) theoretically after infinite time (*1*). Other equations describe the sorption process for different geometries (*2*). This equation indicates that the initial slope of a curve of M_t/M_∞ vs. $t^{1/2}/1$ (a reduced sorption curve), is directly proportional to the diffusion coefficient D. While there are other ways to determine the diffusion coefficient (*1*), this is the technique used in our experiments.

The Diffusion Coefficient

When considering diffusion in polymers, the diffusion coefficient is rarely constant and is, in general, a function of concentration, position within the sample, time, and the stress level within the film. A distinction is usually made between pure concentration dependence in which the diffusion process is still considered Fickian (since Equation 1 applies) and those non-Fickian processes when time or stress dependence appear and whose sorption curves do not fit Equation 1.

Concentration Dependence. By varying the diffusant concentration in a series of absorption experiments, the variation of D with concentration can be determined. This concentration dependence is the normal situation in polymer–organic vapor systems. Since the addition of small molecules to a polymer increase segmental mobility (plasticization), the diffusion coefficient of an organic vapor in a polymer should increase with increasing diffusant concentration (*3*).

Position Dependence. In polymers with heterogeneous structures—for example, semicrystalline polymers and filled elastomers—the transport process is complicated by the generally impermeable dispersed phase. Not only does the crystallite or filler particle create a larger path for the diffusing molecule to traverse, but also the presence of a high area interface within the polymer changes the nature of the continuous phase from that of the pure homogeneous state. These effects are related by the expression (*4*):

$$D = D^*/\tau\beta \tag{2}$$

where D^* = diffusion coefficient in a completely homogeneous polymer, D = diffusion coefficient in the heterogeneous polymer, τ = tortuosity factor, and β = chain immobilization factor.

The tortuosity expresses the average path length a diffusing molecule must travel. In a semicrystalline polymer, it would depend on the crystallinity, crystallite size and shape, and the distribution of crystallites in the amorphous matrix. β is a measure of the effective restraint of the motion of polymer chains in the continuous region exerted by the discrete phase.

Time Dependence. Time dependence may be observed in the diffusion of organic vapors in polymers below their glass transition temperature (T_g) (*5*). At these temperatures, the rate of diffusion is comparable with the rate of motion of the polymer segments. As a result, the value of the diffusion coefficient attained at a given concentration in an element of the polymer will depend on the time for which this concentration has existed at the element. D has more time in which to approach its equilibrium value in thicker films. Therefore, sorption proceeds more rapidly the thicker the film, and the reduced sorption curves do not coincide as required by Equation 1 describing Fickian diffusion.

Stress Dependence. A second non-Fickian anomaly rises because of internal stresses within the polymer, generated during the diffusion process. As sorption proceeds, the unswollen inner core may restrict the swelling outer layer, limiting the diffusion rate in this layer. As a result, it is possible for a thin polymer sheet to absorb more solvent than a thick one in the same time. This has been observed in the diffusion of peracetic acid in styrene–butadiene–styrene (SBS) block copolymers (*6, 7*). Stress dependence can be detected when curves of weight gain per unit area (M_t/A) plotted against the square root of time for different film thicknesses overlap. Crank has carried out detailed mathematical analyses of both time and stress dependence (*8*), providing a way to identify and to distinguish these anomalies.

Case II Transport. At temperatures well below T_g and at penetrant activities near unity, the initial weight gain is proportional to time rather than to $t^{1/2}$. This is termed case II transport (*9, 10*). A boundary exists between swollen gel and glassy polymer, which advances at constant velocity, independent of the sample thickness. The anomaly is believed to arise because diffusion proceeds more rapidly behind the advancing boundary in the gel phase than the polymer relaxations at the boundary itself (*10*). If the penetrant has a sufficiently high activity, the stresses developed at the advancing boundary may be sufficient to cause fracture or crazing of the material.

Vrentas et al. (*11*) have presented a clear analysis of these anomalies based on the Deborah Number, the ratio of a characteristic relaxation time to a characteristic diffusion time.

Applications of Diffusion Analysis

There is an extensive literature demonstrating the effect of polymer structure on the diffusion process (*1*). Only in a relatively few cases has the process been reversed, with the structure of the polymer determined by analysis of the diffusion process.

Diffusion analysis has been used to investigate the effect of aging on the structure of quenched polypropylene (PP) (*12, 13*). Kapur and Rogers (*12*) made diffusion measurements on several films, each having different degrees of aging. The results indicated a significant reduction in diffusivity in the first 30 hr of aging, accompanied by an increase in tensile strength. From these measurements, it was found that in the process of quenching the polymer from the melt, the supercooled liquid collapses into a highly disordered mass of microcrystallites. During the aging process, these microcrystallites fused and became organized. The chain immobilization factor β (Equation 2) becomes considerable, accounting for the rather low diffusion coefficients measured in the aged films. Although reordering of the crystallite tends to make more of the intercrystalline regions accessible, the increased chain immobilization increased the time to permeate those regions resulting in an overall reduction of diffusion coefficient.

The effect of heat treatment on structural order in vinyl chloride polymers was also investigated recently (*14, 15*). Gray and Gilbert (*14*) attempted to determine whether structural changes are likely to occur during the processing of PVC. Their results showed that heat treatment changed the solvent sorption behavior of homopolymers but had no significant effect on the solvent sorption of the PVC/PVAc copolymer. This was attributed to the difference in crystallinity. The copolymer is virtually amorphous and therefore absorbs more solvent and absorbs it more quickly than homopolymers with higher crystallinity. The structure of the low crystallinity copolymer was not affected by heat treatment as observed. On the other hand, the structure of the high crystallinity homopolymer depended on the heat treatment. Quenched samples, in which order would be expected to be minimum, absorbed solvent more rapidly. The sample which was heat-treated below its T_g, and hence would have low structural order, absorbed solvent more quickly than the one heat treated above its T_g. The effect of time of heat treatment was considered to be related to the rate of crystallization.

The effect of mechanical treatment of polymers also has been investigated by diffusion (*16, 17, 18*). The difference between elastic and plastic deformation of low-density polyethylene (PE) film was demonstrated by their difference in transport behavior (*16*). The diffusion constant first increased linearly with strain and later approached a saturation value. This showed that the polymer underwent elastic deformation. The initial increase in transport properties resulted from the increase of fractional free volume as a consequence of the increase of specific volume of the film under tensile strain. After reaching a maximum with increasing strain, the diffusion constant began to drop below the value of unstrained material. This was the result of plastic deformation. The decrease in diffusion constant was caused by an increase in the average path the diffusing molecules had to travel. Such an increase may have been caused within the spherulitic matrix by lamellar rotation, sliding, or fracture, all of which block existing passages through the amorphous component. Eventually, the spherulitic material was transformed into a fibrous structure with different transport properties. The amorphous component in the fibrous material was almost impermeable to solvent. Changes in orientation in polyethylene terephthalate (PET) (*13*) and nylon 6 (*19*) during drawing have also been detected by diffusion measurements.

Recently, Boyer et al. (*2o*) used the diffusion coefficient to detect the presence of a liquid–liquid transition temperature above T_g. Arrhenius-type plots of log D against $1/T$ appear to result in two straight lines. The temperature at the intersection point compares favorably with the T_g determined by other methods.

The application of diffusion analysis to the interphase in styrene–butadiene–styrene (SBS) block copolymers and to the morphology of polyethylene is presented below.

Experimental

SBS. The diffusion coefficient of cyclohexane at 20°C in an SBS block copolymer was measured as a function of film thickness and casting surface. Films of SBS TR-41-2443 (Shell Chemical Co., Houston, TX) were cast under identical conditions from benzene onto glass, Teflon, or previously cleaned mercury surfaces. The films were allowed to dry for a day by evaporation within a partial benzene atmosphere. The films were dried further in a vacuum dessicator for two days before being used in a sorption experiment. The properties of the copolymer are described in Table I.

Electron microscopic observation of the morphology of TR-41-2443 cast in this way shows the expected (*22*) presence of nearly parallel, slightly curved, cylindrical rods of PS in the PBD matrix (*23*).

Polyethylene. The effect of crystallinity on the diffusion coefficient of iso-octane at 54°C in PE was determined in a sorption experiment. Films of Sclair PE (DuPont Chemical Co., Kingston, Ontario) were compression-molded in a hot platens press at 170°C and were allowed to cool in the press. All films were 200 μm thick. The properties of the resins are given in Table II.

Sorption Apparatus. Measurement of diffusion coefficients is simple and can be carried out with a minimum in special equipment. The apparatus used in our investigations is illustrated in Figure 1. A Cahn RG automatic electrobalance was placed inside a sealed chamber and was used to continuously monitor weight gain or loss by the sample. The balance chamber was connected to a solvent reservoir. Before starting a sorption run, the system was completely evacuated. When complete vacuum was achieved, vapor was introduced into the weighing compartment to begin the absorption experiment. After equilibrium was achieved, the system was evacuated to begin the desorption experiment.

Table I. Characterization of SBS TR-41-2443 (*21*)

Block molecular weights	
16,000–85,000–17,000	
Composition	
polystyrene 27.7 wt % (25.1 vol %)	
tri-block 100%	
Polybutadiene microstructure	
cis 1.4	40%
trans 1.4	49%
1,2	11%

Experimental Block Copolymers

Table II. Properties of Sclair Polyethylene Resins

Resin	*8107*	*E341–01*	*8507*	*8707*	*2908*
Plaque density[a]	0.925	0.932	0.939	0.948	0.961
Melt index[a]	4.6	3.0	5.4	5.3	7.1
% Crystallinity[b] (from density)	50	55	59	65	74
% Crystallinity (from DSC)	43	54	61	65	—
Estimated crystal diameter (μm)[c]	50	80	110	140	—
Specific surface area (cm^2/cm^3)[d]	1200	750	550	430	—

[a] Ref. *39*.
[b] $\rho_c = 1.0$, $\rho_a = 0.85$ g/cm (*40*).
[c] From microscopic observation under polarized light (100 ×).
[d] Calculated from crystal diameter, S.S.A. = $6/d$.

Sorption Data. The results were plotted as reduced sorption curves, M_t/M_∞ vs. $t^{1/2}/1$. The apparent diffusion coefficient was calculated to be the average of the absorption and desorption diffusion coefficient:

$$D = \frac{\pi}{32}(I_a^2 + I_d^2) \quad (3)$$

where I_a and I_d are the initial slopes of the reduced absorption and desorption diagrams, respectively. The apparent diffusion coefficient is an estimate of the average diffusion coefficient over the cyclohexane/polymer concentration range between zero and the equilibrium uptake.

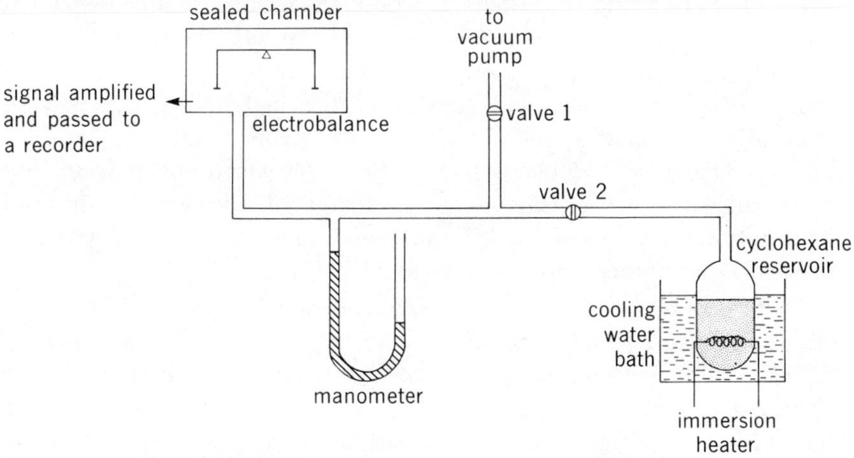

Figure 1. Schematic of the sorption apparatus

Other Analyses. Birefringence measurements using a Babinet compensator were made on the SBS films to detect changes in the orientation of the molecules.

DSC of the PE samples at a program rate of 20°C/min were performed on a DuPont Thermal Analyzer, model 990. A ΔH of fusion = 68.4 cal/g (*24*) for 99% crystalline polyethylene was used to determine crystallinities of the pressed films.

The polyethylene films were viewed under cross polarizers in a Leitz Ortholux microscope with transmitted light to determine the crystallite size.

Results and Discussion

SBS Interphase. Since 20°C is below the θ-temperature for the polystyrene–cyclohexane systems, it was expected that the PBD phase would be permeable to cyclohexane, but the PS domains would be relatively impermeable. (It is known that PS swells almost fourfold in liquid cyclohexane and that SBS may be dissolved even in cyclohexane. However, the maximum uptake of cyclohexane vapor by SBS was approximately 40% of its original weight. Furthermore, a sample of pure PS did not absorb any vapor within the time scale of these experiments. It was concluded then that the pure PS domain was not penetrated by cyclohexane vapor in these experiments and that, except for the interface, the PS domains may be considered an impermeable phase dispersed within a permeable continuum.) Thus the diffusion coefficient would be expected to reflect the structure of the PBD phase and to be characteristic of diffusion in elastomers (i.e., Fickian diffusion).

Unlike conventional rubbers, however, the results are time dependent since reduced sorption curves for films of varying thicknesses do not coincide with each other (Figure 2). This anomaly, which is characteristic of glassy materials, is a sign of the inability of the polymer molecules to respond quickly to the changing concentration. The diffusion coefficient (D) depends on the time for which the polymer and penetrant have been in contact. D has more time in which to approach its equilibrium value in thicker films, and therefore sorption proceeds relatively more quickly the thicker the film.

To account for the apparent restricted mobility of the PBD region attention must be paid to the interphase between pure PS domains and the pure PBD continuum. The presence of a significant interfacial region of mixed composition separating the PS domains from the PBD continuum has been demonstrated by analysis of dynamic viscoelastic behavior (*25, 26*), T_gs (*27, 28*), and small angle x-ray scattering data (*29,*

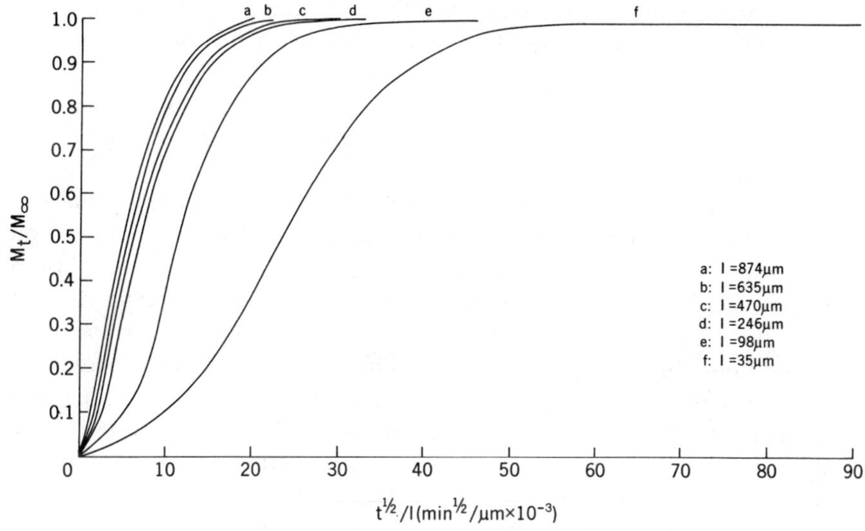

Figure 2. Reduced desorption curves for Teflon cast films of SBS TR-41-2443 of different thicknesses. The absorption curves are similar.

30). The thermodynamic interpretation of Meier (*31*) and Helfand (*32, 33, 34*) have shown the effect of molecular weight and domain shape on interphase thickness and volume fraction. Based on the calculations of Kaelble (*35*), a high interfacial area (about 150 cm^2/cm^3) for similar SBS polymers is expected. This high interfacial area reflects the fact that 3.34×10^{17} domains occupy 1 cm^3 of polymer volume, each with an interphase separating pure PS and pure PBD. According to Meier (*31*), approximately 40% of an SBS polymer with block molecular weight 13,300–75,400–13,300 (slightly shorter than TR-41-2443) is mixed within the interphase, producing a relatively thick interface.

The T_g follows, approximately, the relation for linear additive free volumes (*35*). Thus, the T_g of the interphase decreases monotonically outward from the pure PS domain ($T_g \sim 100°C$) as the volume fraction of PS decreases in succeeding shells to a low of −54°C (T_g of PBD) in the outermost shell. There is then a shell for which the T_g is 20°C, the temperature of these measurements. In fact, at 20°C approximately one-half of the interphase surrounding each domain would be glassy (At the end of the absorption process the glass transition of each shell would be less because of the plasticization effect of the absorbed cyclohexane. Nevertheless, at the beginning of diffusion and until T_g became 20°C, the diffusion process would exhibit non-Fickian characteristics.).

Absorption of cyclohexane by this sizeable interphase fraction would be expected to show non-Fickian characteristics caused by slow molecular motion at temperatures below the T_g. Below the T_g the polymer molecules are effectively immobile, and hence the rate of approach to the equilibrium conformation upon absorption is hindered, and time-dependent characteristics appear.

Internal Stress in SBS Films. In these same experiments, the approach to equilibrium of the apparent diffusion coefficient depended on the casting surface used in preparing the SBS films. The equilibrium diffusion coefficient is the diffusion coefficient corresponding to the equilibrium conformation of polymer. It was obtained by extrapolating the apparent diffusion coefficient to infinite film thickness where time dependence should disappear. It also is an estimate of the average diffusion coefficient over the concentration range studied, but in a polymer with equilibrium conformation. Furthermore, the equilibrium conformation represents an average conformation over the range of concentration from zero to the equilibrium uptake.

Figure 3 shows the approach to equilibrium of D against film thickness for different casting surfaces and film pretreatments. It can be seen that the approach to equilibrium of the diffusion coefficient is quicker in the intermediate thickness range for the mercury cast films and

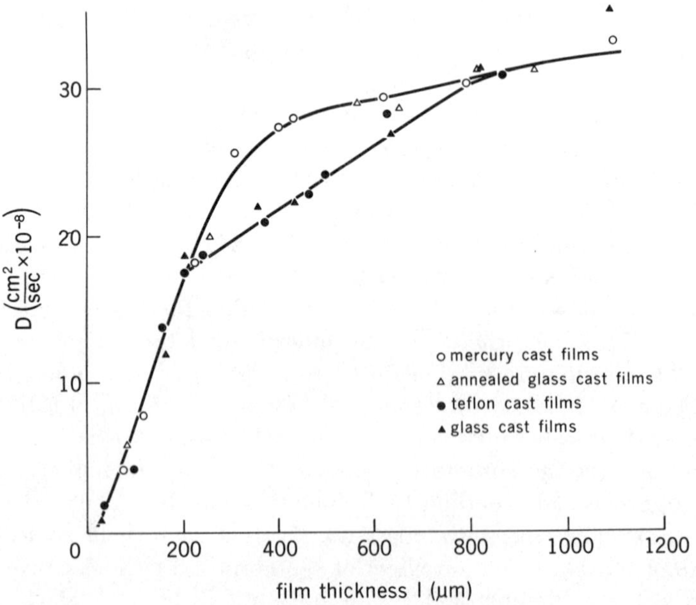

Figure 3. Effect of film thickness on the apparent diffusion coefficient defined by Equation 3.

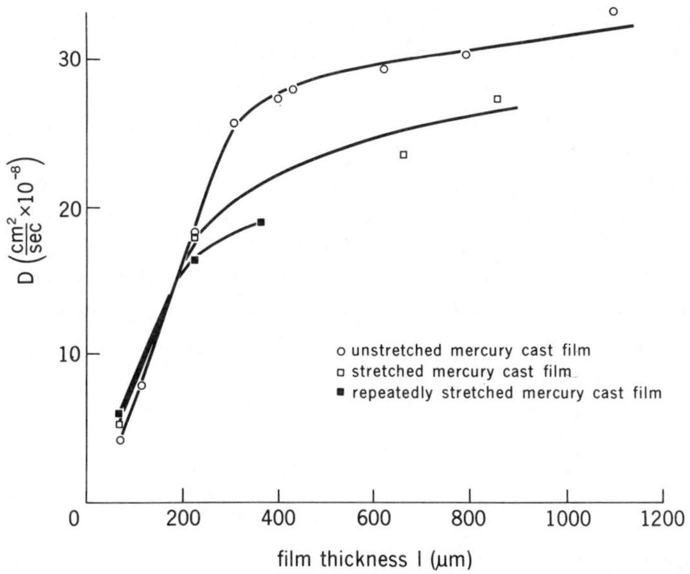

Figure 4. Effect of film thickness on the apparent diffusion coefficient defined by Equation 3 for mercury cast films.

annealed glass cast films. The glass cast films were annealed by heating a freely suspended film at 110°C for 15 min. These films were assumed to have an equilibrium structure and not to be under internal stress. The glass and Teflon cast films remained under stress because of adhesion between the annealed films and the casting surface during drying of the film.

It is well known (*36*) that stress can induce chain orientation along the direction parallel to the applied stress. As a consequence of internal stresses within the glass and Teflon cast films, the high degree of orientation of polymer chains favors closer packing than is possible in a completely randomized amorphous polymer. This results in a higher density, and hence the fractional free volume is smaller than in the equilibrium amorphous state. Such an effect is reflected in the decreased rate of equilibrium approach of the diffusion coefficient for the glass and Teflon cast films observed in Figure 3.

To demonstrate this hypothesis a little further, the mercury cast films were deliberately stretched approximately 15–20% during the sorption experiments. The results, when compared with unstretched mercury cast films (Figure 4), show an expected reduction in diffusion coefficient. The same films were then allowed to relax for seven days and were stretched again to the same extent as previously for a repeat cycle of sorption experiments. The results show a further drop in diffusion coefficient indicating possibly even greater chain orientation. This additional orientation may have been induced before a complete recovery of the polymer chains was possible. Slow recovery of the polymer chains after stress was relieved, was presumed to result from restricted mobility caused by the interphase.

There was no change in physical properties such as birefringence or density which would correspond to the more pronounced change in diffusion coefficients. This indicates the far greater sensitivity of transport properties to chain orientation. Furthermore, this indicates the essentially surface nature of the diffusion measurements. The observed internal

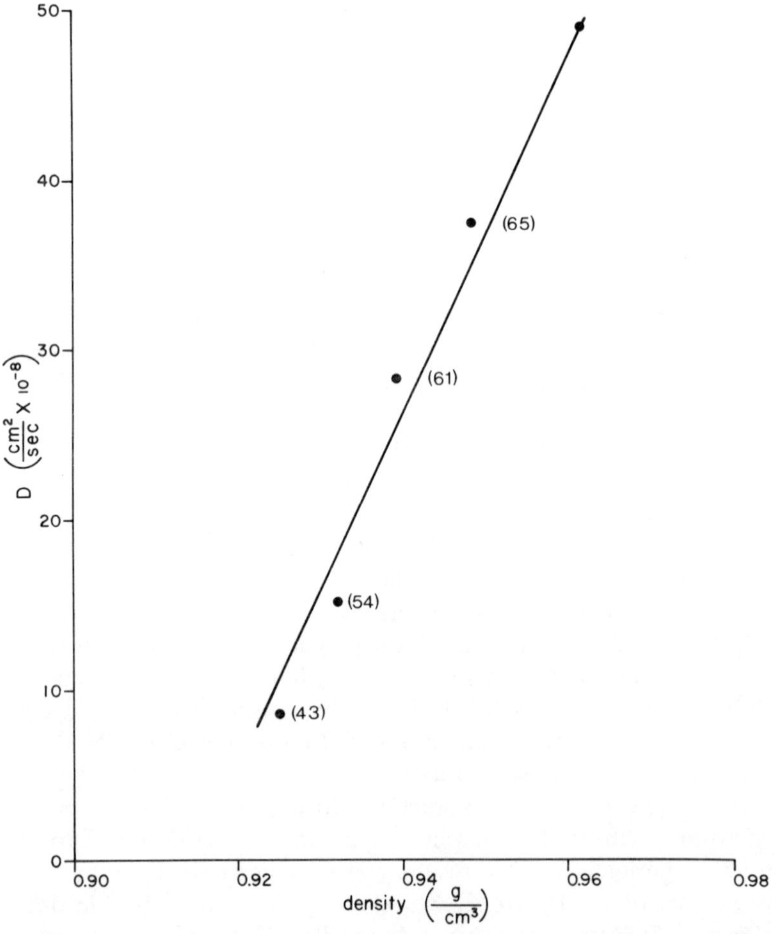

Figure 5. Effect of crystallinity on the diffusion coefficient of iso-octane in Sclair PE. The numbers in parentheses are the percent crystallinities by DSC.

stresses were the result of changes in the SBS surface effected by the casting surface. It is suggested that these and other surface structures also could be determined in larger, finished products, providing a means to investigate the effect of processing conditions on the surface of the polymer product.

The equilibrium diffusion coefficient D_e for glass and Teflon cast films is identical with that of mercury cast and annealed films. This, in fact, should be expected since after stress is relieved from glass and Teflon cast films upon absorption, the polymer molecules return to their normal orientation as in the mercury cast and annealed films. Therefore, the equilibrium diffusion coefficient is the same regardless of the casting surface.

Polyethylene. A particular difficulty in the use of diffusion measurements to analyze structure is the interpretation of the results. Unlike other structural analysis tools which give a direct measure of a particular structural feature, the diffusion coefficient is indirectly related to structure. Any given change in diffusion coefficient may be the result of any one of several different structural changes. The diffusion measurements must generally be interpreted in conjunction with other information.

In an attempt to confirm that changes in diffusion coefficient in well known systems are consistent with expected behavior, a series of PE's with varying density were investigated. As expected the fractional uptake of solvent decreased with increasing film density or crystallinity. However, surprisingly, the diffusion coefficient increased with increasing density or crystallinity (Figure 5).

It appears that in these Sclair resins the increased crystallinity is the result of fewer, larger crystallites rather than more of the same size. These larger crystallites have a lower total surface area. The tortuosity is then relatively unaffected, but the chain immobilization factor β defined by Equation 2 is reduced significantly, resulting in a larger diffusion coefficient in the denser, more crystalline materials.

Previous measurements of diffusion in polyethylene (*37, 38*) have found a diffusion coefficient decreasing with increasing crystallinity. These results were explained by assuming an increased tortuosity with increased crystallinity. These investigators have compared samples of polyethylene prepared by different manufacturers under different conditions. For their samples the increase in crystallinity may be the result of more crystallites of a smaller size, increasing both the tortuosity and the chain immobilization factor.

The Sclair resins under study here were of similar melt index and were manufactured under similar conditions. They are a uniform series of polyethylenes differing only in density. Microscopic observation under polarized light confirmed the hypothesis that the higher crystallinity in

Sclair resins results from fewer, larger crystallites. These results are given in Table II along with estimates of the specific surface area to which β is assumed proportional. Thus even in relatively well defined systems, diffusion measurements provide new information on polymer structure.

Conclusions

The measurement of the diffusion coefficient is an excellent tool for investigating subtle changes in polymer structure. Both the interphase in styrene–butadiene–styrene block copolymers and the morphology of polyethylene have been investigated using this method of diffusion analysis.

Despite the difficulties of interpretation of these measurements and of identification of the various anomalies, the diffusion coefficient is a valuable parameter which provides structural information at a level that no other technique can provide. Furthermore, because diffusion proceeds inwards from a polymer surface, the surface is the prime structural unit studied by this technique. Diffusion analysis is expected to be an important tool for investigating the effect of processing on polymer surface structure.

Acknowledgments

The authors are grateful to the National Research Council of Canada for the financial support for this work. Thanks are also due to the Shell Chemical Co., Houston and the DuPont Chemical Co., Kingston for providing sample polymers for the research.

Literature Cited

1. "Diffusion in Polymers," J. Crank and G. S. Park, Eds., Academic, London, 1969.
2. Crank, J., "The Mathematics of Diffusion," Oxford, London, 1956.
3. Fujita, H., *Fortschr. Hochpolym.-Forsch.* (1961) **3**, 1.
4. Michaels, A. S., Vieth, W. R., Bixler, H. J., *J. Appl. Polym. Sci.* (1964) **8**, 2735.
5. Park, G. S., in "Diffusion in Polymers," J. Crank and G. S., Park, Eds., Academic, London, 1969.
6. Sefton, M. V., Merrill, E. W., *J. Polym. Sci., Polym. Chem. Ed.* (1976) **14**, 1581.
7. Ibid. (1976) **14**, 1829.
8. Crank, J., *J. Polym. Sci.* (1953) **11**, 151.
9. Alfrey, T., Gurnee, E. F., Lloyd, W. G., *J. Polym. Sci.* (1966) **12**, 249.
10. Hopfenberg, H. B., Frisch, H. L., *J. Polym. Sci.* (1969) **B7**, 405.
11. Vrentas, J. S., Jarzebski, C. M., Duda, J. L., *AIChE.J.* (1975) **21**, 894.
12. Kapur, S., Rogers, C. E., *J. Polym. Sci.* A-2 (1972) **10**, 2107.
13. Blackadder, D. A., Vincent, P. I., *Polymer* (1974) **15**, 2.
14. Gray, S., Gilbert, M., *J. Polym. Sci., B* (1975) **16**, 387.

15. Illers, K. H., *Makromol. Chem.* (1969) **27**, 1.
16. Yasuda, H., Peterlin, A., *J. Appl. Polym. Sci.* (1974) **18**, 531.
17. Williams, J. L., Peterlin, A., *J. Polym. Sci.* A-2 (1971) **9**, 1483.
18. Warwicker, J. O., *Br. Polym. J.* (1972) **3**, 68.
19. Takagi, Y., *J. Appl. Polym. Sci.* (1965) **9**, 3887.
20. Boyer, R. F., Enns, J. B., *Polym. Prepr., Am. Chem. Soc., Div. Polym. Chem.* (1977) **18**(2), 462.
21. Keelen, T. L., "Experimental Block Copolymers," Technical Information, Shell Chemical Co. (1972).
22. Soen, T., Inoue, T., Miyoshi, K., Kawai, H., *J. Polym. Sci.* A-2 (1972) **10**, 1757.
23. Sefton, M. V., Sc.D. thesis, Massachusetts Institute of Technology (1974).
24. Wunderlich, B., Cormier, C. M., *J. Polym. Sci.*, A-2 (1967) **5**, 987.
25. Shen, M., Kaelble, D. H., *J. Polym. Sci.* (1970) **B8**, 149.
26. Kraus, G., Rollman, K. W., *J. Polym. Sci., Polym. Phys. Ed.* (1976) **14**, 1133.
27. Bares, J., *Macromolecules* (1975) **8**, 244.
28. Kumler, P. L., Keinath, S. E., Boyer, R. F., *Polym. Prepr., Am. Chem. Soc., Div. Polym. Chem.* (1977) **18**(1), 313.
29. Hashimoto, T., Nagatoshi, K., Todo, A., Hasegawa, H., Kawai, H., *Macromolecules* (1974) **7**, 364.
30. Hashimoto, T., Todo, A., Itoi, H., Kawai, H., *Polym. Prepr., Am. Chem. Soc., Div. Polym. Chem.* (1977) **18**(1), 295.
31. Meier, D. J., *Polym. Prepr., Am. Chem. Soc., Div. Polym. Chem.* (1974) **15**(1), 171.
32. Helfand, E., *Acc. Chem. Res.* (1975) **8**, 295.
33. Helfand, E., Wasserman, Z. R., *Macromolecules* (1976) **9**, 879.
34. Helfand, E., Wasserman, Z., *Polym. Prepr., Am. Chem. Soc., Div. Polym. Chem.* (1977) **18**(1), 290.
35. Kaeble, D. H., *Trans. Soc. Rheology* (1971) **15**, 235.
36. Peterlin, A., *J. Macromol. Sci., Phys.* (1975) **B11**, 57.
37. Rogers, C. E., Stannett, V., Szwarc, M., *J. Polym. Sci.* (1960) **45**, 61.
38. McCall, D. W., *J. Polym. Sci.* (1957) **26**, 151.
39. Wiseman, K. D., personal communication (1977).
40. Van Krevelen, D. W., "Properties of Polymers," 1972.

RECEIVED November 10, 1977.

INDEX

INDEX

I

K

L

M

N

Q

R

S

The text of this book is set in 10 point Caledonia with two points of leading. The chapter numerals are set in 30 point Garamond; the chapter titles are set in 18 point Garamond Bold.

The book is printed offset on Text White Opaque 50-pound. The cover is Joanna Book Binding blue linen.

Jacket design by Sharri Harris. Editing and production by Candace A. Deren.

The book was composed by Service Composition Co., Baltimore, Md., printed and bound by The Maple Press Co., York, PA.